Geological Excursions in Lakeland

Geological Excursions
in Lakeland

by

E. H. Shackleton, F.G.S.

DALESMAN BOOKS

1975

The Dalesman Publishing Company Ltd.,
Clapham (via Lancaster), North Yorkshire.

First published 1975

ISBN: 0 85206 292 3

The front cover photograph of Stickle Tarn and Pavey Ark is by Christopher Hanson-Smith. The back cover photograph shows the Scafell range with Upper Eskdale in the foreground. All unacknowledged photographs and diagrams in the text are by the author.

Photographs are on pages 65-68 and 85-88.

Printed by Galava Printing Company Limited,
Hallam Road, Nelson, Lancashire.

Contents

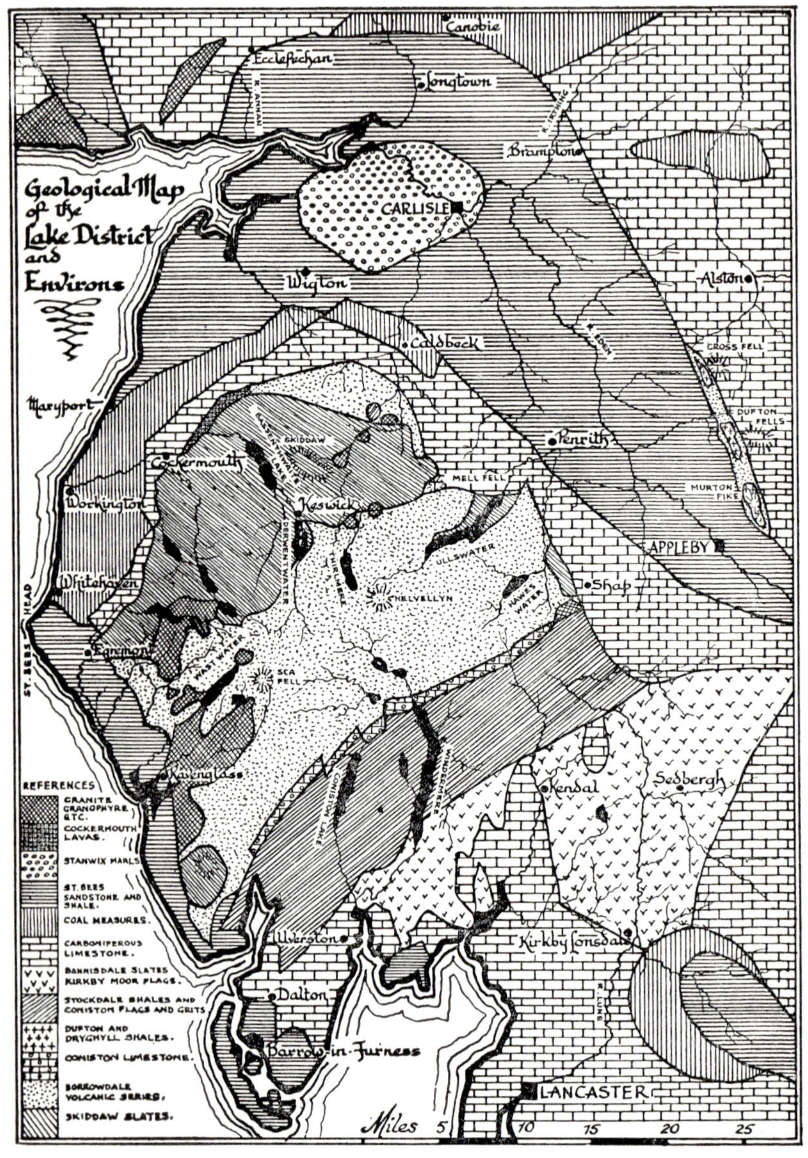

—map drawn by E. Jeffrey.

Preface

I HAVE always believed that some knowledge of the history of any science and its development can be of assistance in gaining a full understanding, and certain it is that a knowledge of the story of our Lakeland pioneers adds a good deal of interest, and often understanding, in the complicated field of our Lakeland geology. That the story does not go all that far back in time in no way detracts from it, for from its earliest days it has been filled with the most interesting and often quite prolonged controversies. To this I must be careful to add that some of these arguments still go on, and in attempting to give guidance I can only do so in the light of my own knowledge and experience of the geology of our fascinating district over many, many years.

I often say "A yard of seeing is worth a mile of hearing"—a memory perhaps of my own young days when I was constantly abjured to "go and see !"—so in the following pages, wherever possible, I shall endeavour to guide my reader's footsteps to places where whatever is being discussed can be seen. So far as I am able the excursion form will be used, but I shall take the liberty of mentioning other locations which are worth a visit, even though they do not fit so neatly into the excursion pattern. Given such locations, and where to find them, the reader can then visit them at his, or her, convenience.

In mentioning geological locations no automatic right of access is implied, and it is always advisable, wherever possible, to obtain permission from the owners. Finally may I make a plea? To the good geologist his hammer is a tool, sometimes a very necessary one let us admit, but it must always be used with a great deal of discretion. To hammer everything in sight, just for the sake of hammering, is at once to proclaim the utter amateur! It must be remembered that the number of people who take an interest in our rocks, and the science of geology, grows year by year, and it becomes increasingly important that we should preserve what we have. In illustration may I mention an exposure where a piece of obvious Skiddaw Slate could be seen embedded in the Purple Breccia. Recently someone has removed it! To what end? If a piece of Skiddaw Slate was required Skiddaw is made of it.

Even if a specimen showing the slate in breccia was required I could show you where plenty of pieces lie around, but this was in situ, and once removed from its setting this bit of slate loses all significance. The action has simply robbed countless other interested people of the pleasure of seeing it, so please, please use your hammer with the utmost discretion, and leave as much as possible for other visitors to enjoy.

N.B. To avoid confusion between historical and present day references, the traditional names of Cumberland and Westmorland have been used throughout this book and no attempt has been made to refer to regions of the new county of Cumbria.

Acknowledgements

My major debt is undoubtedly to the many great men of the past, who not only carried their hammers for years throughout our lovely Lakeland hills and dales meticulously observing and noting the facts, but subsequently recorded what they had seen in language which all who knew something of their science could understand. To many of my past students, friends and members of our Cumberland Geological Society who have helped with suggestions and advice, field work and photographs, my sincere thanks and appreciation. To all the present day workers mentioned in the text whose work I have followed and reported with interest, many thanks.

For help and criticism of the MSS my special thanks to Dennis E. Jackson, Ph.D., B.Sc., F.G.S.; to A. P. Rook, B.Sc., F.G.S.; and to H. A. White, B.A. I must not forget the Librarians, both past and present, of the Geological Society who have been so helpful time after time in providing copies of geological papers, both present day and more often long out of print.

We all, knowingly or unknowingly, stand upon the shoulders of those who have gone before and if my modest effort can be taken as a tribute, however small, to past Lakeland "knights of the hammer" and as encouragement to those who would follow in their footsteps I shall be well content.

E. H. Shackleton, F.G.S., 1975.

1.   Early Days—The Story of the Skiddaw Slates

THAT great series of rocks to which Skiddaw mountain gives its name is undoubtedly the oldest in our Lakeland, but for a very long time controversy has raged about the true age and the succession, and even today not all geologists who know them are in complete agreement. In attempting to give the history of our knowledge of the Skiddaw Slates one cannot ignore the work of that gifted amateur Jonathan Otley, who in working out for the first time the succession of our local rocks originally called them simply "the Lower Slates". In this he was contrasting them with the not greatly dissimilar rocks in appearance, the Silurians of southern Lakeland—"the Upper Slates, with a limestone at their base". This limestone was of course the Coniston, and during our wanderings we shall certainly come back to it—but first a word about Jonathan.

Jonathan Otley was born on Loughrigg near Ambleside on the 11th of February 1766. His father, who was by trade a basket maker, seems to have been a well educated man. He was a good Latin scholar and he encouraged his son in his studies under the parson, the Reverend R. Steele, in his school in Langdale, and later at school in Ambleside. Perhaps I may point out that Lakelanders have for long paid great attention to education. Many of our schools, humble though they may be, go far back into history, like the one at Crosthwaite, Keswick, which even in the days of Elizabeth I was spoken of as "our school . . . time out of mind". After he left school Jonathan worked with his father at Scroggs making baskets until he was 25, and during this time he became very skilful in the art of cleaning and repairing watches and clocks as well as in the art of engraving. In 1791 he moved to Keswick and set up in business as a watch and clock repairer, living for some five years at Browtop. He then moved down into town and took a room through the arch from the Moot Hall and up a series of steps, whence he became known as "Jonathan Otley up t' steps". Today there is a little memorial tablet on the wall commemorating him as geologist and watchmaker.

Otley's great interest in our mountains and the rocks from

which they are carved caused him, over the years, to be more and more sought after for his practical guidance and fund of information by such eminent men as Airy the Astronomer, John Dalton the chemist and not least by Adam Sedgwick who, in the early 1800s, was beginning the attempt to unravel the tangle of our older rocks. Otley in 1823 first produced his now famous map and guide to the Lakes, and in it he divided up into three great series our local rocks. Jonathan it was who first gave a description of what might rightly be called the lithology of the Skiddaw Slates, and their relationship to the Skiddaw Granite upon which they lie. If there was some trace of the Wernerite ideas still current at the time, Otley's own observations disposed him more towards treating the granite he so meticulously described as an intrusion, and he makes mention not only of the alterations in the slates around the granite but also of the minerals to be found like moylbdena, apatite and wolfram! In all this our early geologist showed a remarkable grasp of the rocks of our area and their distribution. "Of these divisions", he says, "the first in the series forms Skiddaw, Saddleback, Grasmoor and Grisedale Pike, with the mountains of Thornthwaite and Newlands; it extends across Crummock Lake, and by the foot of Ennerdale as far as Dent Hill, and after being lost for several miles, it is elevated again at Black Combe."

If this is a good illustration of Otley's grasp of the distribution of our rocks on the grand scale his observations in detail are no less meticulous, and must always be a shining example to all who would become "knights of the hammer". It must be remembered that it was Otley who first drew attention to the phenomenon of cleavage in rocks, and perhaps this is a good place to discuss it and to answer a query I have often had. Why slates, since the Skiddaw variety, as compared to say the slates of Borrowdale, seem to have such an indifferent cleavage? Here I think we are dealing with the difference between the layman and the geologist. As Jonathan first pointed out a "slate", in the geological sense, is the result of a process which produces a very definite alteration in the structure of the rock, enabling it to be cleaved **across** the bedding to give man good thin roofing material. This involves not only a re-arrangement of the particles of the rock but often the growth of new minerals which themselves are easily cleaved. Where the rocks of Skiddaw have been used for roofing, close examination will show that you are dealing with rocks that have split **along** the bedding planes. It may well be that in places the cleavage coincides with the bedding, but in fact they are thinly bedded siltstones, which you can see exposed on the fellside between Mungrisedale and Mosedale. Hereabouts there are still old buildings so roofed and my guess is that this practice goes back to long before geology as a science was even thought about.

The term "slate" as referred to the rocks of Skiddaw was a layman's and not a geologist's; after all they were "slating" roofs way back in Tudor times.

After Jonathan Otley and his great friend, Adam Sedgwick, who was the first to make mention of fossils in our Skiddaw Slates, came Harkness and Nicholson, both of whom had Lakeland connections. Separately and together these two tackled the slates and made collections of fossils, mostly graptolites. Although that remarkable man, Henry Alleyne Nicholson, was born in Penrith, he was for a time Professor at a Canadian University where he extended his study of graptolites to the strata of North America. Professor Nicholson was, in fact, very much of a pioneer in the then new science of palaeontology, the study of fossils. On his return to his native land it was he who first drew attention to similarities between the graptolites of our Skiddaw Slates and those of the Lévis Shales of Quebec.

And what is this long standing controversy all about? Really it is one of relative age and just where our slates should fit in the geological age scheme. Our earliest workers were not bothered too much about it because in their day only two systems of rocks were recognised low down in the Palaeozoics (the rocks that were typified by the remains of ancient life forms)—the Cambrian and the Silurian.

The oldest rocks which contain fossils in Wales were called by Adam Sedgwick the Cambrian, after Cambria the ancient name for Wales, and these rocks were divided from the unfossiliferous rocks, the Pre-Cambrian, by a great unconformity. From this starting point Sedgwick worked his way up through younger and younger strata, collecting and noting the increasingly numerous fossils, and often making comparisons with our rocks of Skiddaw and their fossils. His great friend and fellow worker, Roderick Impey Murchison, began his researches by working **down** from the base of the Devonian system along the Welsh Borders, and to these newly described beds he gave the name Silurian (again after the name of a Welsh tribe). For long the two workers continued their researches blissfully ignorant of the fact that they were applying different names to at least some of the beds that were identical! Unfortunately this state of affairs continued for so long that when at last it was realised what had happened both had become eminent men of science, with positions to defend, and much bitterness ensued between the two erstwhile friends. Eventually, using graptolites as markers, Charles Lapworth, himself at the time an amateur geologist, ended the argument by taking some of Sedgwick's higher beds and some of Murchison's lower ones, and founding a completely new system, the Ordovician, and this again was an old tribal name in Wales. It was this refinement, the introduction of this new system, which intensified controversy here in Lakeland where Lapworth's graptolite scheme proved to be useful.

From this time on the zoning, that is the division into sets of beds that could be identified by their fossil content, went on apace until in many parts of Britain most of the strata from the base of the Cambrian upwards were, more or less, readily identifiable. But what of our Lake District? In the latter part of the last century our rocks too received increasing attention. Miss Gertrude L. Elles and her co-worker, Miss E. M. R. Wood, produced monographic works on the graptolites. Elles identified four zones of Tremadocian (then regarded as Upper Cambrian) and Arenignian age in 1898, and by 1932 had considerably refined this scheme. Her succession, based on our Skiddaw Slates, became the standard Arenignian graptolite succession for Britain. Certain of these zonal graptolites she insisted were of Tremadocian age, thus implying that at least some of our Skiddaw Slates were Cambrian, but no one, not even Miss Elles herself, ever regarded the rocks of Skiddaw as simple. However, it was this suggestion that part at least of the Slates were of Cambrian age that intensified controversy.

With this argument still unsolved the next workers to come upon the scene were the officers of the Geological Survey when, in the early 1920s, they found that the geology of the Whitehaven-Workington area included large stretches of Skiddaw Slate country. E. E. L. Dixon in 1924 arranged the succession of the Slates in this way:

> (d) Mosser Slates (with Watch Hill Grit)
> (c) Loweswater Flags
> (b) Kirkstile Slates
> (a) Blakefell Mudstones

This arrangement was based largely on lithology (i.e. the nature and physical composition of the rocks). Fossils were not of frequent occurrence, it was said, and there was little attempt made at this time to utilise them. May I say that attempting to sort out a succession by tracing beds that look alike (i.e. by their lithology) has for long been and to some extent still is a source of trouble. It cannot be too strongly stressed that whilst lithology can often tell us much about conditions of deposition, source and direction of origin as well as, sometimes, the right way up, it can say little that is in any way reliable about the relative age of strata. Given the same, or a similar source of material, and the same conditions of deposition you will end up with the same lithological type of rock, irrespective of the time of formation. Another thing which for long was not clearly realised was that

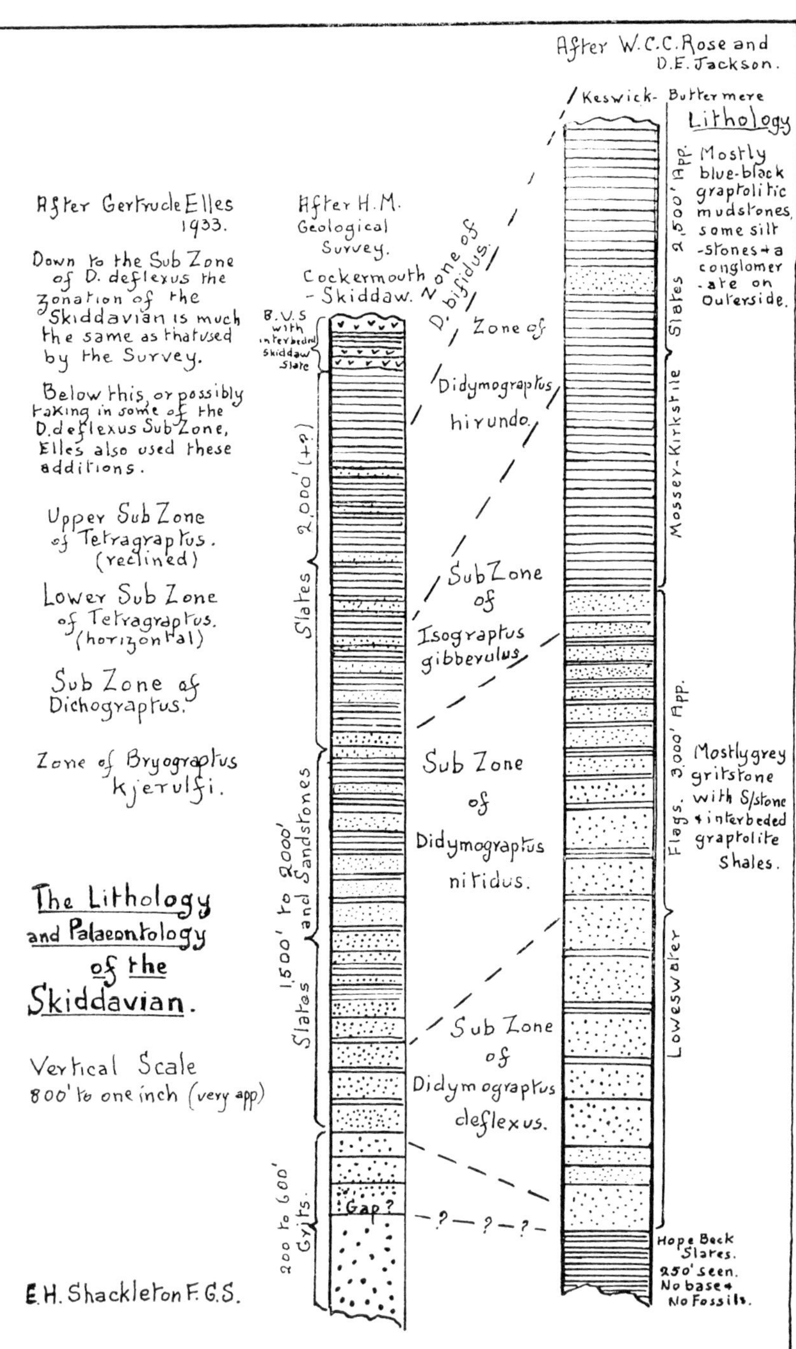
After W.C.C. Rose and D.E. Jackson.
Keswick - Buttermere
Lithology
Mostly blue-black graptolitic mudstones, some silt-stones + a conglomer-ate on Outerside.
2,500' App.
Mosser-Kirkstile Slates
After Gertrude Elles 1933.
Down to the Sub Zone of D. deflexus the zonation of the Skiddavian is much the same as that used by the Survey.
Below this, or possibly taking in some of the D. deflexus Sub Zone, Elles also used these additions.
Upper Sub Zone of Tetragraptus. (reclined)
Lower Sub Zone of Tetragraptus. (horizontal)
Sub Zone of Dichograptus.
Zone of Bryograptus kjerulfi.
The Lithology and Palaeontology of the Skiddavian.
Vertical Scale 800' to one inch (very app)
E.H. Shackleton F.G.S.
After H.M. Geological Survey.
Cockermouth - Skiddaw.
Zone of D. bifidus.
B.V.S with interbedded Skiddaw Slate
Zone of Didymograptus hirundo.
Slates 2,000' (+?)
Sub Zone of Isograptus gibberulus
Sub Zone of Didymograptus nitidus.
Slates 1,500' to 2000' and Sandstones
Sub Zone of Didymograptus deflexus.
200 to 600' Gjits
Gap?
— ? — ? — ?
Loweswater Flags. 3,000' App.
Mostly grey gritstone with S/stone + interbeded graptolite Shales.
Hope Beck Slates. 250' seen. No base + No Fossils.

large areas of our slates, even well away from Skiddaw, had been affected by thermal metamorphism. Eventually the Survey officers themselves realised that Blake Fell Mudstone was largely Kirkstile Slate that had been so affected, for we find them saying that the two rocks pass imperceptibly into one another.

After the work on the Whitehaven-Workington area, which resulted in one of the finest Memoirs of its time, the Survey moved over to the Cockermouth district and began work in that wild stretch of country from Cockermouth eastwards across Skiddaw itself. Unfortunately the second World War intervened and not until recently has the result of all this work been published. Meanwhile, a geologist born and bred on the Skiddaw Slates has produced, what is to me, the most critical and definitive account to date. Dennis E. Jackson was born in Cumberland and went to the Cockermouth Grammar School where, under Mr. D. Rigby, he received his introduction to geology. Then followed a degree at King's College, Durham University, and a Ph.D. thesis on the country lying between Cockermouth and Mungrisedale, almost totally Skiddaw Slate country.

I think it is only fair to say that Jackson started with a very great advantage—he was a native! While the advantages of this may not be at once apparent to an "offcomer", believe me they are very real! You don't have to learn the geography, you have already played for years over the ground and, whilst the mention of some obscure location by a local would bring little response to others, the native knows at once where to go! Then the "Skids" are the "Skids" and much that is quite obvious to the native would have to be patiently learned by others. Having said this I can only pay tribute to the meticulous and careful way in which Jackson worked, for over this very rough area he carried out a most exhaustive examination of the Skiddaw Slates basing his work, not exclusively but very largely, on the fossils they contained. Every location mentioned by Elles, and indeed earlier workers, he methodically checked, collecting from them at least 1,500 identifiable graptolites. To anyone who has ever tried serious collecting from our Skiddaw Slates this is a remarkable achievement in itself. The specimens were all identified by Professor Bulman and the collection has a place of honour over on the North-East.

Jackson and the Survey officers, notably W. C. C. Rose, all found that not all the zones erected by Elles could be maintained and I have attempted, by means of a composite diagram, to bring together the results of all these long years of work. I gather from Jackson and Rose (who covered much of the same ground) that, while there are still some small differences between them, there is also a large measure of agreement. Of all the zones erected by the late Gertrude Elles those put forward on the compound forms *Bryograptus* and *Dichograptus* proved to be least satisfactory and are now considered by many workers to have little or no value. But it was on the occurrence of these very forms that argument had for so long been centred! It now seems that in our local rocks these compound forms have a very long range in time, and odd specimens can be expected to turn up in beds the fossils of which are far more representative of much younger zones. This, as was pointed out by various workers long ago, is very much the case in the Lévis Shales of Quebec and I have myself found on the Dodd a very fine *Dichograptus* from a location most notable for the prevalence of *Phyllograptus.* Another source of difficulty is the fact that our Skiddaw graptolites are most easily found and collected from screes—fair enough, but in many cases this leaves at least some doubt as to the exact horizon to which the finds should be referred.

One very good example of this is Barf (218265—referred to as "The Bishop" on the $2\frac{1}{2}$ in. O.S. map). Barf is the well-known hill near the Swan Hotel at Thornthwaite from the screes of which many graptolites have been collected over the years. At one time or another I think I have seen every type of graptolite found in the Skiddaw Slates, from *Bryograptus* to a somewhat dubious *Didymograptus protobifidus,* from this locality. Mind I am not claiming I found them! One has to remember that many such finds by others are brought to me for identification. Be this as it may it does raise a question to which we so far have no fully satisfactory answer. One suggestion was put forward by Gertrude Elles herself, and that was what is known as a condensed succession. This implies that here a much less thickness of strata accumulated in the same time that much greater thicknesses were laid down elsewhere. Discussing this with Jackson recently he discounted the idea, pointing out that the lithological type is much the same at Barf as it is elsewhere, whereas a condensed succession would require a change to much more finely grained rocks. His tentative suggestion is that down the south-eastern side of the mountain the rocks are stacked in great folds, that these rocks have been heavily attacked by the frosts of our recent Ice Age, and that this has given rise to heavy weathering from mixed horizons, the debris accumulating on the screes from which practically all our finds have come. He feels this area could very well prove a rewarding one for some patient fossil hunter willing to give his time to it. To this I can only add that a close perusal of any work such as this can only show time and again how much our science is indebted to the dedicated amateur. Usually it is someone having both ready access, the knowledge, the time and the determination, who digs out the result so long sought in vain by others! So with these remarks and suggestions, and leaving the experts to wrangle over the technical details, perhaps we might now go and see something of these intriguing—if exasperating—Skiddaw Slates for ourselves.

Excursions:

A.   Skiddaw Dodd

I THINK that a useful introduction to the Skiddaw Slates and their many complexities might be made by visiting some of the exposures along the Forestry roads of Skiddaw Dodd. This is the conical peak which stands out from the southern face of the mountain itself above the valley and lake of Bassenthwaite. Leaving Keswick by the Carlisle road and crossing the railway bridge, pass by the road end on your right that leads to Applethwaite, noting on the valley floor to your left the rounded mounds, somewhat elongated along the valley, which are drumlins left by the ice. Before long you will come to what is obviously a new stretch of road with a lovely greenstone wall on the right. This example of the waller's art is built from volcanic ash and is worth a good look as it is the product, not of our local Honister Quarries but of the upper band of slate from Broughton Moor above Coniston. A little further along part of the old road has been left by a farm, while opposite is a fine exposure of the slate we wish to see (240266—Long Close).

This is a really good exposure and worth examining. For the most part the slate is of the dark, well bedded type which often is fossiliferous and should be carefully examined for such remains, but I must confess that although I scarcely ever pass without a look I have so far had no luck! However, if there is little to see beyond a slate exposure along most of it, this in itself is useful for only by seeing such exposures over and over again can one confidently recognise the slates when one encounters them! But this does not exhaust interest here for towards the far end of the rock wall there is a sudden change in the appearance of the rock itself. In fact there is here an igneous intrusion in the form of a dyke and this is a very good place to "get your eye in". As between sedimentary rocks (rocks laid down in water, of which our Skiddaw Slates are an example) and igneous rocks (rocks that have reached their present site in a molten form), there is a distinctly different look which to the trained eye is immediately obvious. One has to admit that to the beginner this may be very far from obvious, so this is a good place to gain a little experience.

If you break a piece of this grey looking rock off, you will find it rather fine grained and exhibiting little structure. Ward called it a mica trap, a term which has now become obsolete, and it would probably be described today as a felsite. But as with all exposures it is best to stand back and look at it as a whole. Often there is far more to be learned in this way than by contenting oneself with chipping a bit off and using a lens. Here you can see wnat a dyke looks like, and most important, you can note for future reference the difference between the two types of rock, sedimentary and igneous.

Continue now to a little beyond the fourth milestone from Keswick, and here on the forest side there is now a new car park and the beginning of Forest Trails on the Dodd. If you have plenty of time these may well be worth following as they are sure to give access to the forest roads, all of which expose the Skiddaw Slates. However, to begin our own particular trail follow the road a little further until, with the private woods of Mirehouse on the left, a road runs up into the forest from a lay-by on the right (235283). At about seventy paces from the road there is a gate, sometimes open, sometimes not, and I think its real purpose is to keep motor cars from invading the forest roads. A little way above this gate there is a wide clearing with one road going straight on and another making an abrupt turn left. Take the sharp turn left and walk slowly up the hill.

Walking up this road, which has been recently widened, towards the next clearing the Slates are exposed most of the way on the right. Again they have the dark even appearance which suggests fossils so you might keep a good look out. Notice too that there are pieces of broken rock lying about with stringers of white quartz running through. This is a fault breccia, and was turned out of the bank when the road was being cut, for hereabouts there are many faults. At the next clearing there is a steep rock face as you come up the rise. Again one road goes straight on while the other doubles back on itself to run uphill. This rock breast is well worth attention for it is upon an actual line of fault. Both on the rock face itself, and lying about on the ground, there is much fault breccia. If you have never seen a fault breccia before, examination will show it to be composed of broken angular fragments cemented together, and the cement here upon the Dodd is usually quartz. First notice that the fragments *are* angular for this is the feature that distinguishes all breccias, whatever their origin. Usually a rock composed of re-cemented broken fragments naturally contains many little cavities. These are known as "vugs" or "druses", and here such vugs often contain quite nice little crystals of quartz. The two terms, "vug" and "druse", are really miners' terms and in the mineral veins of our local mines the spaces can be of considerable size. Their importance is this: before crystals can form there must be space into which they can grow, and here you can see on a small scale the kind of location from which come the many beautiful mineral specimens on display in the Keswick Museum.

Although this is a fault face slickensides are not much in evidence, for the rock has tended to break up rather than to slip and polish the line of break. Over to the right of the exposure the flexuring of the beds can be seen, while low down on the extreme right a small exposure shows ripple marks. This can be taken as an indication that the water in which the rock was laid down was not too deep and that the detritus from which the rock was formed was silty (i.e. it contained some sand) rather than pure mud, for mudstones without this sandy element rarely show ripple marks.

Proceeding up the hill the rocks are well exposed most of the way on the left, and again, where not broken by faulting, are the kind of slates which usually yield fossils. But there is ample evidence that we are still following a line of fault and in about 275 paces you will come to a steep bluff with really magnificent folding and faulting. Running up this breast the rocks rise at a high angle and display some very fine folds: then, it would seem the rocks have broken, for running almost vertically downwards they stand on end. Away to the right, and higher up, the slates run in small tightly compressed folds. This exposure is a splendid example of the folding and contorting the Skiddaw Slates have suffered through the pressure of earth movements over the long, long ages that have passed since they were first laid down and it seems almost incredible that rock, quite hard rock, can be so bent and folded. What one must remember, however, is that such folding movements take time, often a very long time indeed.

Examine this exposure with care: photograph it by all means for it is a splendid example of folding, but please do remember that this is a case where your hammer is not required. No one who has any claim to being a geologist would want to disturb this exposure in any way; it must be left for others to see. Now continue for about 70 paces to where a streamcourse comes down on the left to pass beneath the road. Notice that the slates exposed above the stream seem rather hard and somewhat bleached and that a few feet above these slates the character of the rock alters. If you have really benefited from examining the exposure on the main road you will readily recognise a similar exposure here, for this is another of the felsite dykes which are quite a feature of the Dodd. Indeed, it has been suggested that it is the prevalence of these dykes that gives the hill its conical form. The dyke here is badly weathered and the brown colour is due to the iron content of the rock. I am afraid that this

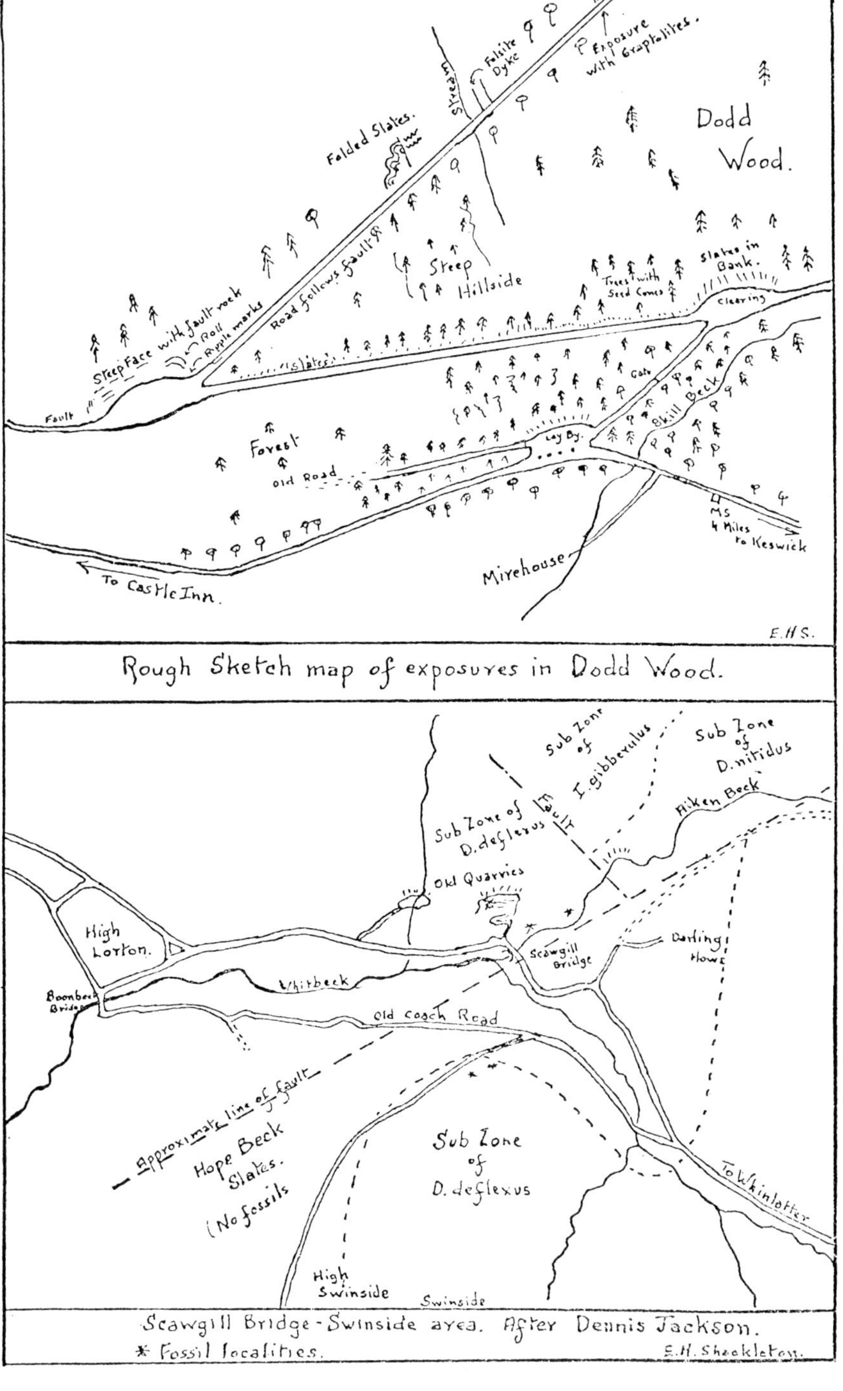

Rough Sketch map of exposures in Dodd Wood.

Scawgill Bridge - Swinside area. After Dennis Jackson.
* Fossil localities.
E. H. Shackleton.

weathering has penetrated pretty deeply and getting fresh rock is almost impossible.

Now continue uphill for another hundred paces or so to yet another exposure in the bank. Usually it is all too obvious that this is a location of great interest for often rock has been dug out of the bank and left lying about. In fact this is a very fossiliferous exposure and many really fine graptolites have been found here. *Didymograptus* occurs only sparingly, but *Phyllograptus* (both species), *angustifolius* (the long type) and *typus* (the shorter one) can be found in an excellent state of preservation. Once, and once only so far, I have found that rather rare graptolite which gives its name, rather incongruously, to the zone, *Isograptus gibberulus*.

I know it is beginning to get rather difficult to extract fresh material from the bank, and this is all the more reason why such pieces that are extracted should be examined with the utmost care, then split and examined again to make sure there are no fossils before disposing of them. Frequently when I have visited this exposure I have found perfectly good fossils on rock people have dug out and then left lying around! Fossils, good fossils, in our Skiddaw Slates are all too rare to be treated like this. Besides, it is becoming of increasing importance to remember that there are other interested people besides yourself. When you have finished fossil collecting here please do not go away leaving the debris strewn around for someone else to tidy up. It is a privilege to come into this lovely forest, so please don't spoil things for others!

All around these roads on the Dodd there are reminders that the area is much broken up by faults and along these breaks in the rock there has been much infiltration of quartz. For some unexplained reason mineralisation on the southern aspect of Skiddaw seems to be confined to quartz. In the slates themselves you may have noticed peculiar square holes, usually about one quarter of an inch across. These are what is left after the cubic crystals of iron pyrites have weathered away. The mineral iron pyrites, or "fools gold" as it is sometimes called, is a sulphide of iron. This is readily attacked by the elements and changes to the sulphate which is easily soluble. If you break up one of the pieces which shows the square holes you can often get out perfect little cubes of the iron pyrites. The presence of much iron pyrites in a mudstone is usually an indication of bad bottom conditions at the time of deposition—a lack of oxygen and the occurrence in quantity of sulphuretted hydrogen, conditions which are inimical to most life. On the other hand, since graptolites were free floating creatures rather than bottom dwellers, such conditions can hardly be held responsible for the scarcity of their remains. There is another sedimentary structure which can often be found in the darker Skiddaw Slates of the Dodd. It is known as "cone in cone" structure and this admirably describes it, for a specimen does look as if a series of cones has been fitted neatly inside one another. The cause of this interesting structure has baffled geologists for a long time and even that greatest of authorities on sedimentation, Twenhofel, while offering several possibilities, will not commit himself.

As the forest roads are on private ground it is as well to contact the Head Forester first by ringing Bassenthwaite Lake 957/272. Forest Trails help, but such permission gives greater freedom and the Dodd and its roads provide a happy hunting ground as well as many lovely vistas.

Excursions:

B. Whinlatter and Scaw Bridge

ANOTHER excursion which at least has the advantage of ready accessibility is the area traversed by the Whinlatter Pass. Leaving Braithwaite by the Lorton road ascend the pass to the new car park (224245), with its magnificent views across to Skiddaw and down Bassenthwaite to Binsey. At the head of the lake is a low lying alluvial plain laid down since the last retreat of the ice some ten to fifteen thousand years ago. Near the lake there are still rather muddy, reed-covered areas which are often inundated after rain, but the fields further back are already drained and cultivated.

Notice that at the western end of the car park a large boulder stands by the side of the road. I have often looked rather fearfully for this interesting boulder, half expecting it to have disappeared in some misguided "tidying up" operation. So far, thank goodness, it has survived! Let us take a closer look for it is worth inspection. To those already well acquainted with our local rocks this is obviously a block of Borrowdale Volcanic Breccia, but what is it doing here? A little searching around will show many rock exposures nearby but none of them is of volcanic origin, they are all exposures of Skiddaw Slate. So this boulder is an erratic (i.e. a piece of rock out of context, so to speak). It just does not belong here: it was carried up by the ice from Borrowdale and left here as the ice melted, some 650 feet above sea level, to provide an excellent example of an erratic and a "perched block".

Now let us walk a few yards down the hill to the small quarry on the right known as Knott Head. There are roadside exposures near the quarry—all are Mosser-Kirkstile Slates in the zone of *Didymograptus extensus,* a subzone of *Isograptus gibberulus*— and here we can look for fossils with some hope of success. At various times I have found myself and helped others to find quite a few fossils here, many of them in an excellent state of preservation. But first a few words about looking for fossils. Don't expect them to give themselves up! They always need looking for, and in my experience it does take a little time to "get one's eye in". The best way to find fossils is to find one! Having found a specimen you then know exactly what to look for, and having found your first you can often go on to find others. As these rocks are in the zone of *Didymograptus extensus* the most likely find is this graptolite or, as we are here near the top of the zone, a variant *Didymograptus nitidus.* Fairly commonly too I have found *Azygograptus,* a single stiped form with sharply triangular thecae. *Tetragraptus* too I have found, but the type fossil of the zone, *Isograptus,* I have never yet found here. Perhaps this is not too surprising as Jackson tells me that on the whole it is far from being a common fossil.

Perhaps a few words of explanation about fossil zones and zonal fossils would be worthwhile. A fossil zone is a thickness of rocks which can be recognised by the fossil assembly it contains (as a rule not by a single fossil but by an assembly of fossils, and ideally the thickness should not be too great). To be really useful for the purpose of zoning the fossils that mark a zone should (a) have a wide distribution, (b) have a short range in time, and (c) be fairly easily recognised. To fulfil all these requirements it will be readily understood that not just any kind of fossil will do, and it must be admitted that in the Skiddaw Slates such fossils are by no means as common as we would like. It was at one time hoped, if not taken for granted, that these long extinct forms, being free-floating, would have a very wide, even world wide, distribution and might thus serve the purpose of a world wide cross-correlation of strata. I understand from Jackson who has done much work on graptolites of the American continent that some workers now begin to doubt this, while from a recent paper read at a meeting of the Geological Society others are claiming a very close resemblance between Alaskan graptolites and our own in upper Ordovician times.

When you have tired of Knott Head you can hopefully try many of the exposures nearby for almost all have yielded specimens. I have even found *Phyllograptus* from the small exposure under the wall immediately opposite the quarry. Making your way in the direction of the head of the pass there are many other exposures before you come to a guest house by the roadside. Beyond this the road dips before it rises to the top of the pass

Zonal Graptolites of the Skiddaw Slates.

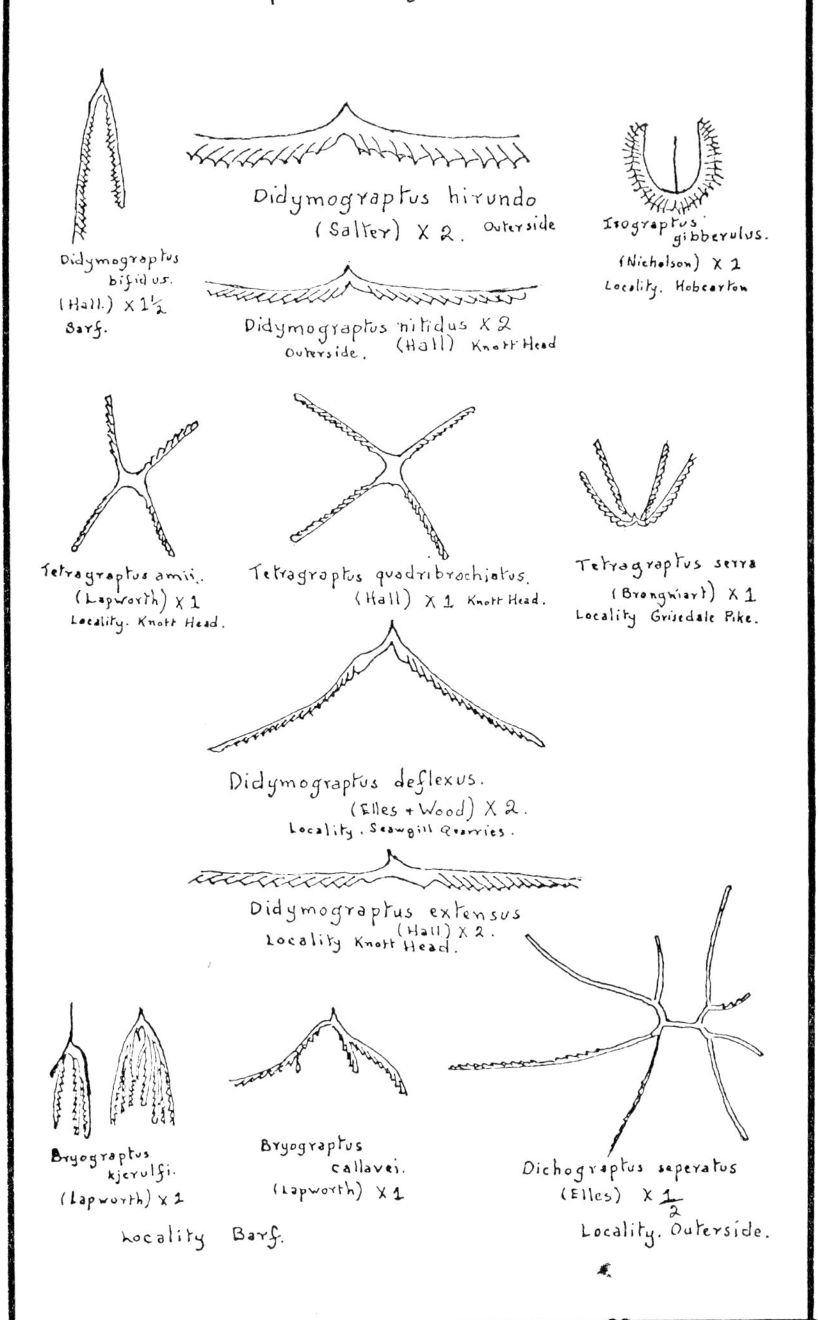

Didymograptus bifidus.
(Hall.) X 1½
Barf.

Didymograptus hirundo
(Salter) X 2. Outerside

Didymograptus nitidus X 2
Outerside. (Hall) Knott Head

Isograptus gibberulus.
(Nicholson) X 1
Locality. Hobcarton

Tetragraptus amii.
(Lapworth) X 1
Locality. Knott Head.

Tetragraptus quadribrachiatus.
(Hall) X 1 Knott Head.

Tetragraptus serra
(Brongniart) X 1
Locality Grisedale Pike.

Didymograptus deflexus.
(Elles + Wood) X 2.
Locality. Seawgill Quarries.

Didymograptus extensus
(Hall) X 2.
Locality Knott Head.

Bryograptus kjerulfi.
(Lapworth) X 1

Bryograptus callavei.
(Lapworth) X 1

Dichograptus seperatus
(Elles) X 1½
Locality. Outerside.

Locality Barf.

and near here there is a forest road on the left. Until recently this was a Forest Trail but as this seems now to have been discontinued perhaps it will be as well to seek permission. On this hillside, where tree felling has been under way of late, there are many roads and exposures and while I have not made a detailed examination I have found specimens of both *Didymograptus* and *Phyllograptus.*

Continuing in the direction of Lorton you can look out for two foresters' cottages standing a little way back from the road on the left. Behind these cottages you can look up one of Lakeland's most magnificent coombs, Hobcarton Gill. Like most of our coombs it has a northern aspect, and thus sheltered to some extent from the sun a small upland glacier continued to exist long after our major valley glaciers had disappeared. That its life was indeed long this great coomb itself bears testimony, for it is nearly a mile in length and something like three-quarters of a mile wide, while at its head it is one thousand feet deep. To produce such a great coomb enormous quantities of Skiddaw Slate must have been removed and, even allowing for the fact that much of our slate is fairly soft, it is still a great reminder of the erosive power of ice.

Continuing on your way in the direction of Lorton you must make a steep descent to Scawgill Bridge and the nearby Scawgill Quarries. This area is not, geologically, a simple one for it is cut by faults, and so rocks of several different horizons crop out as shown by the rough sketch map on page 19. However, for fossils it is fairly good and the many small quarries, besides the main one, are worth attention. All the quarries on the hillside to the north of the stream are in the zone of *Didymograptus deflexus* and the rock, the Loweswater Flags. Here you can expect to find the type fossil and some of its variants for, besides *D. extensus, Didymograptus-v-fractus* and *Didymograptus nicholsoni* have been found.

A footpath leaves the road by a stile near the bridge and leads to a waterfall known as Spout Force. All along this path there are exposures and small quarries, and here I have more than once found the type fossil of the lowest sub-zone of the Skiddaw Slates, *Didymograptus deflexus.* In the slates exposed near the waterfall, having crossed one of the local faults, you are in the *Didymograptus nitidus* sub-zone and might try your luck for this fossil. Both these sub-zones are in the Loweswater Flags.

Back on the road by the bridge, and before pressing on towards Lorton, look back towards the hill down which you came and notice the deeply cut gorge by the side of the road. This has been excavated along one of the local faults and is a good illustration of the effect of such breaks in the strata on scenery. In about half a mile in the direction of Lorton you will find a side road on the left. If you go down here quietly, for it is steep and narrow, you will pass the doors of some little cottages before coming to a stream by the road side. In the far bank of this stream slates are exposed, and these, the Hope Beck Slates, are the oldest of the Skiddaw Slates. So far no fossils have been found, and while there are exposures on the south bank of Whitbeck all are poor. Perhaps the best place to look would be along the beck (164240) which gives the slates their name. Who knows? Often it **is** the amateur who finds the fossil the experts have sought for years in vain!

Take care in coming out of this lane by car for you will emerge onto a major road. Follow this after turning left, until the main road turns right while the minor road goes straight on. Crossing Boonbeck Bridge make your way carefully up what is almost a one track road back towards Whinlatter. This is the old coach road over the pass; at the point where it begins to descend and a road bears off to the right (179254), leave the car on the wide verge. Following this road to the right there are two locations of interest; one is a quarry quite near the road, the other exposure being much further over, almost on the skyline. I must admit that I myself have never been too successful hereabouts but I mention the location because of the unusually wide variety of graptolites reported by Jackson in his paper "Graptolite Zones in the Skiddaw Slate Group of Cumberland". Both of these exposures are just within the *deflexus* zone of the Loweswater Flags; the type fossil is said to be common, while we also have mentioned both *D. nitidus* and *D. nicholsoni.*

In making these two excursions we have seen quite a thickness of the Skiddaw Slate Group, from the Hope Beck Slates to the Kirkstile Slates, and while the expert would probably have started with the lowest beds exposed and worked his way up, for us perhaps the interest of the hunt in such lovely settings will be enough. Maybe it was noticed that I did not give a zonal reference for the exposures on the Dodd. This was because I have nowhere seen a firm statement as to the zones here, and I believe that the south face of Skiddaw is still awaiting the attention of the experts. However, the very common occurrence of *Phyllograptus* has always predisposed me to think of the *Isograptus* zone, but never until recently had I found this fossil there.

Now for a few other interesting locations. The zone of *Didymograptus nitidus* is exposed in a quarry in Stonycroft Gill (228212). This location is worth a visit, graptolites apart, for it was here that the German miners built a dam in the 1560s to divert the water from the bed of the stream and the mineral vein that followed it. It is also here that one of the last smelt mills to be operated in this area was situated. There is still much slag about and the colour of some of it seems to suggest the smelting of cobalt from the mine up on Scarr Crags. Much of it however was from the smelting of lead, as even lead from Greenside was at one time smelted here. The rough road leads on to Scarr Crags, and on Outerside there are several locations for *Didymograptus hirundo* while even *D. proto-bifidus* has been reported. This is the earliest of the "tuning-fork" graptolites; another good area for this zone, and the *D. bifidus* type, is along the old road that follows the shoulder of Southerfell from Scales to Mungrisedale. There are interesting exposures near the farm of Hazlehurst and several along the road towards the hamlet.

With this we must leave our interesting, if somewhat exasperating Skiddaw Slates, just expressing the hope that you may make finds for yourself. If Dennis Jackson could find over 1,500 identifiable graptolites from these same Skiddaw Slates there is little need to stress that they can be found, however elusive they might seem!

References

Elles, Gertrude L., 1898. "Graptolite Faunas of the Skiddaw Slates", Q.J. Geol. Soc., Vol. 54, pp 463-539.

Elles, Gertrude L., 1933. "Lower Ordovician Graptolite Faunas with special reference to the Skiddaw Slates", Summary of Progress for 1932, Geol. Survey pt. 2, pp 74-111.

Harkness, R., 1863. "On the Skiddaw Slate Series", Q.J. Geol. Soc., Vol. 19, pp 113-140.

Harkness, R., 1863. "On the Fossils of the Skiddaw Slates", Reprint British Association, Geol. Section p 69.

Jackson, Dennis E., 1961. "Stratigraphy of the Skiddaw Group between Buttermere and Mungrisedale", Geol. Mag., Vol. 98, pp 515-528.

Jackson, Dennis E., 1962. "Graptolite Zones in the Skiddaw Group", Journal of Palaeontology, Vol. 36, No. 2.

Nicholson, H. A., 1868. "The Graptolites of the Skiddaw Series", Q.J. Geol. Soc., Vol. 24, pp 125-245.

Rose, W. C. C., 1954. "The sequence and structure of the Skiddaw Slates in the Keswick-Buttermere area", Proc. Geol. Soc., Vol. 65, pp 403-406.

The collection of rocks, fossils and minerals as well as the library of geological works that belonged to Professor Harkness is now in the keeping of the Carlisle Museum at Tullie House, Carlisle, whilst some of Gertrude Elles's original graptolites are in the Keswick Museum.

2. The Borrowdale Volcanics
and their relationship to other rocks

OF all the problems posed by Lakeland geology perhaps one of the most persistent, and certainly one of the most hotly debated, has been the succession between the volcanics and the other rocks, both above and below. In 1868 H. A. Nicholson, while studying medicine at Edinburgh, produced as his Doctorate thesis an "Essay on the Geology of Cumberland and Westmoreland". For this he was awarded a gold medal and in it he clearly summed up the then state of our knowledge.

In his introduction he says: ". . . the base of the Lower Silurians is formed by the Skiddaw Slates, a vast series of sedimentary rocks with a thickness of some 7,000 feet . . . representing the base of the Lower Llandeilo, or Arenig Group. The Skiddaw Slates are *conformably succeeded* by the Green Slates and Porphyries . . ." The italics are mine, but note first "Lower Silurians" for what is now called the Ordovician, and notice also how correct was his assessment of the stratigraphic position as Arenig. His estimate of the thickness of the Slates is in line with all but one of our most recent workers. Not only did he regard the succession between the Skiddaw Slates and the Borrowdales as conformable, but the same, he wrote, applied to the upper junction between the Borrowdales and the Coniston Limestone, and it is these two junctions that have been so hotly debated ever since!

The old N.N.E./S.S.W. faults he described as running parallel to the great Pennine Faults, but there were also strong faults which ran almost at right angles to them and these, he held, were of later date. The series of rocks was very strongly folded, and this folding brought up the Skiddaw Slates in great anticlines. Sedgwick too held that the lower junction was a conformable one, but he did express doubt about the one between the Borrowdales and the overlying Coniston Limestone.

However, in 1869 Nicholson changed his mind about the lower junction, although to some extent this was foreshadowed in his Chapter I, "The Green Slates & Porphyries". In introducing his subject and acknowledging his great debt to Professor Sedgwick ". . . to whom we owe most of our knowledge of this group", he goes on to say ". . . whether the Green Slates and Porphyries are really conformable with the Skiddaw Slates is a question which admits of some doubt . . ."

Apparently these doubts had grown and what caused him to change his mind was his investigation of a locality not far from his own home at Penrith. This location is in a steep wooded gill that comes down the mountainside above the Howtown shore of Ullswater. Nicholson called it Egg Beck, but if you want to check for yourself you had better try and avoid the trap into which I fell. Probably during the "romantic period" when such prosaic names as Egg Beck were out, the name was changed. Fortunately, after some puzzlement, I met one of the older inhabitants who told me with a grin, "Aye . . . aye lad, I ken Egg Beck awlreet . . . but they call it Aikbeck these days". So, off on the right foot I found my beck (470232) above the new caravan site at Parkfoot, close by the new works for a pumping station for Manchester Corporation's Waterworks. The gill is much overgrown but following it up I found closely alternating beds of fine black shale (with well preserved graptolites) and thin beds of sandstone or siltstone. The dips, I noticed, were very variable and at one place in the bed of the gill there was a perfect little anticline. Above the trees there is an exposure in the bank of what is called "mottled shale" and this is overlain, a few feet away but with no actual junction visible, by Borrowdale Volcanics. Now it is noticeable that the dips (i.e. the slope of the beds) are dissimilar, so a geologist would say that the two beds had incongruent dips, suggesting an unconformity. And this is exactly the conclusion drawn by Henry Alleyne Nicholson! Over the next few years he spent much time on the investigation of this junction between the Skiddaw Slates and the succeeding Borrowdales, and over one hundred years ago he published in the Quarterly Journal of the Geological Society his findings of such an investigation stretching over the country between Ullswater and Keswick. He concluded that all the junctions which he found exposed were faults.

In 1872 The Survey Memoir covering the area was published but added little to our knowledge. Aveline, who was the District Officer at the time, said when questioned that the only unfaulted junction between the two series was on the fellside above a place called Bootle in West Cumberland. In 1876 John Clifton Ward's *Geology of the Northern Part of the Lake District* appeared, only to add in some respects more confusion to what was already becoming a very confused scene indeed. Ward put forward the idea that most of the porphyries and amygdaloids (i.e. the lavas) were in fact highly altered and metamorphosed tuffs and ashes, and went on to suggest that the granitoid masses were themselves but an end point in this process of intense metamorphism.

If the latter part of this statement foreshadowed some present day thinking, and was of course long before its time, the part deriving lavas from ashes could not be sustained, and in the mapping of that complex series of rocks—now beginning to be known as the Borrowdale Volcanics rather than the Green Slates and Porphyries—it did much to confuse and very little to help. On the other hand if you ever try to follow any of Clifton Ward's maps, as I have done myself, you will find them a fine example of geological field work, incorporating much careful and meticulous observation; you will, however, have to make allowance for his own particular terminology. To a large extent the Survey seemed to accept Ward's interpretations and included them in the new maps. On these maps *all* the junctions we are at present discussing were shown as faults. The Whicham Valley strike fault (of which more hereafter), first mentioned by Adam Sedgwick, was discarded, as was the one along the length of Haweswater.

Ward seems always to have regarded the B.V.S.-Coniston Limestone junction as an unconformity and, reading again this fine old work, it is pretty clear to me that, however much later workers have fastened upon his lower faulted junctions, Ward himself was in no doubt that *originally* this junction was quite normal. On page 69 he says: "It is very unfortunate that in the whole of this area there are no clear sections showing an unfaulted junction between the Skiddaw Slates and the Borrowdale Volcanics." He then goes on to remind his readers that he has already drawn attention to an exposure in a small gill just south of the summit of Eycott Hill where a bed of slate six to ten feet thick is interbedded with the volcanics. He mentions Aveline's junction on Black Coomb, and that he and Dakyns have seen volcanics interbedded with Skiddaw Slates in the area west of Shap which he goes on to describe as a "complete passage". From a number of other references made it is evident that he himself had little doubt that, whatever he had recorded on his maps as the present state of the junction, originally there was no break. Indeed on page 70 he states this specifically: "The close of the Skiddaw Slate period was marked, not by any upheaval and consequent denudation, but by a gradual coming in of volcanic action over the old sea bed." In the long years that have followed it is really quite remarkable how geologist after geologist has picked upon the faults Ward so faithfully recorded on his maps, and yet totally ignored so clear a statement of what was, in essence, the crux of the matter.

In an anniversary address to the Geological Society in 1891 Sir Archibald Geikie, giving a general revue of volcanic activity in the Lake District, first suggested that much of this activity was of a submarine character. Around this time too Harker and

Marr were engaged on work on the Borrowdales with an increasing emphasis on their petrology and in this they were greatly assisted by W. M. Hutchins. Perhaps it is not inappropriate to add that Ward had himself been one of the first to make this approach, no less than two dozen beautifully coloured microscopic sections adorning the first pages of his book. What these must have cost him in time and labour only those who have ever tried making a rock section by hand will appreciate! However, it was in 1892 that J. E. Marr, in an article in the Geological Magazine, described the junction between the Borrowdales and the Coniston Limestone as a faulted one; this was the first suggestion of such a fault. But things were getting really lively about this time when a visiting Scandinavian geologist declared that the Borrowdales were obviously Pre-Cambrian!

In 1897 the Survey Memoir on the Appleby-Ullswater district came out and proclaimed that ". . . as elsewhere the rocks of the Volcanic Series were faulted against the Skiddaw Slates . . . these faults, however, were of the low angled variety, what are now called thrusts". There was one little helpful gleam of light about this time, for W. W. Watts, perusing the petrology of the rocks, clearly showed that some of Ward's "altered ashes" were demonstrably lavas.

In 1900 J. E. Marr gave the views of himself and his co-worker, Alfred Harker, about the structure of the Lake District. The old rocks of the area, he maintained, had been pushed over towards the north by a great force acting from the south. Not only had the rocks moved, but they had moved at different rates relative to each other, with the Borrowdales tending to lag behind the Skiddaw Slates, and the upper Silurians lagging in turn. Thus horizontal fissures had been formed along which the movements had taken place: these fissures he called "lag faults".

Thus since the 1860s our Lakeland geology had undergone a change almost as savage as that which had affected the rocks themselves! The Borrowdale Volcanics, both at their base and again at the top of the series, were now separated from the fossiliferous beds by great "lag-planes" or "thrusts". A really tremendous thrust, it was now suggested, underlay the whole district to bring the Skiddaw Slates over the Coniston Limestone at Drygill. The Whicham Valley and other strike faults had completely disappeared. But it must not be imagined that these somewhat extravagant theories gained ready or complete acceptance. Geikie in the first volume of his *Ancient Volcanoes of Britain* gave an excellent revue of the Borrowdales and stated (as a footnote on page 229), ". . . the large number of faults said to separate the Skiddaw Slates from the Borrowdales could not be proved, and probably did not exist". J. D. Kendall would have none of the "lag faults" either. He showed from his work in the

Ireleth district that the older rocks had been both faulted and eroded before the deposition of the Coniston Limestone.

Professor Marr repeated his views on the structure of the Lake District, and its effect on the scenery, in an anniversary address to the Geological Society in 1906 and persisted in them when, in 1910, the Geologists' Association published its well known *Geology in the Field.* I suggest that one must try and remember the geological climate of the time when attempting to assess the pros and cons of this great controversy. It was about this time that the great thrust planes discovered in the Alps and in the North-Western Highlands were being avidly discussed, so there seemed little reason why we, in Lakeland, should not have them too! Professor Marr, at the head of the Cambridge School of geologists, and backed to some extent by at least some members of the Survey, constituted a powerful body of opinion indeed, and perhaps it is not surprising that for quite a time they went unchallenged. But reading again many of these old papers on our local geology, I have been repeatedly surprised to find how often there has been a tendency to elevate theory into fact and to go out and find not the facts so much as the facts that fit the theory! This is a very unscientific approach and I am a little apprehensive that we are not altogether free from it today.

However, only two years after the publication of *Geology in the Field* two papers were read at the Geological Society, both on the same area, which were to herald yet another period of fierce debate. First came a paper by Bernard Smith on glaciation in the Black Coomb area, but in it he almost incidentally restored the Whicham Valley Fault. This was followed by a paper from a complete amateur, John Frederick Norman Green, on "The Older Palaeozoics of the Duddon Estuary". Both papers were read in 1912, but Green's paper the Society refused to publish in its Proceedings or Quarterly Journal, so Green published the paper himself at his own expense. Perhaps we might look at this paper in at least some detail if only to try and discover why the Society took so very unusual a course.

Green, as I have said, was an amateur geologist. By profession he was a civil servant at the Colonial Office, so his field work was to a large extent a spare time occupation. However, his opening move was such that he could undertake the work at home, or at any rate no further away than the offices of the Geological Survey, where he studied in great detail the meticulous mapping of the old survey officers on their original six inch sheets. His careful study of the maps of the Whicham Valley area led him to have grave doubts about the interpretation currently being put upon them, and the succession, both between the Skiddaw Slates and the Borrowdales, and the Borrowdales and the Coniston Limestone. To Green the area seemed admirably suited to his purpose. It was not too difficult of access, it was fairly compact with little faulting or folding, and it was relatively free from the complications often introduced by much igneous intrusion. Altogether it seemed admirably suited for a re-investigation into the structural relationships between these rock groups.

Tackling first the Skiddaw Slate/Borrowdale junction he quite clearly established to his own satisfaction that there was a gradual and uninterrupted passage between the two series of rocks. It was in fact possible, at places like Baystone Bank, to obtain a hand specimen which showed "blue shale" (as the Skiddaw Slates were called locally) at the bottom of the specimen passing upwards into what was known as "mottled shale", and this, Green showed, was simply consolidated Skiddaw mud with stringers and blotches of volcanic ejectamenta. This "mottled shale" was traceable upwards in many places into either a volcanic ash or lava of the normal Borrowdale type. At the upper end of the succession he was able to demonstrate that, as Sedgwick had first suggested, the Coniston Limestone lay uncon-formably upon various members of the Borrowdale succession.

Thus began a single-handed and courageous assault in this area on problems, the complexities of which had daunted geologists, professional and amateur alike, and over the next ten years he tackled one area in the Lake District after another. Long years of work by later geologists have not always upheld his work in detail for, in trying to carry the succession he first established in the area betwixt Millom and the Duddon into other areas, he tended to an over-simplification. Be this as it may, anyone interested, and wishing to get some understanding of this vexed problem of succession, would be well advised to consider his work and views.

Having said all this, and before we betake ourselves off to look for some of this evidence, perhaps it would be as well to consider in a little detail just what is implied by the terms "conformable", "unconformable", "conformity" and "unconformity". When rocks are quite clearly bedded, and one bed is seen to be lying snugly upon the one next below without any divergence, such an expo-sure would usually be regarded as showing a "conformable junction", thus implying that throughout the time when these beds were being deposited there had been no break in deposition. Now if we think for a moment about how and where such beds are laid down it becomes apparent that if we were to follow them far enough we should reach a place where, at the time of this deposition, there was a shoreline along the body of water in which the beds were being laid down, and here there may well be exposures which show rocks of a very different kind. Even where deposition has been continuous and the beds are conformable it need not mean that all the beds are of the same character, for during the time of deposition the depth of water may have changed considerably, strong currents may perhaps have swept in, and the kind of rock formed changed too. Nevertheless, deposition has clearly gone on all the time.

What is implied by an unconformity more definitely than anything else is an age gap, for what it means is that a series of deposits were laid down and subsequently consolidated into rock. At some time after this induration the rocks were subjected to the pressure of earth movements that altered their position from the near horizontal beds which were laid down to a more steeply inclined, or even to a much folded condition. During, or after these movements, the rocks became exposed to the forces of denudation, and they were attacked and worn down until a land surface was formed across their eroded edges. Then the land sank again (or the sea rose) and a fresh set of beds were deposited, the bedding planes of which bore no relation whatever to those of the older rocks. Since all this would take time, often a very considerable time, it may well be that the remains of once living things preserved in the two respective strata will bear little or no relationship either, depending very largely upon the relative rate of evolution of living things at the time.

All degrees of discordance are met with in the field even when dealing with a true unconformity, depending very largely upon the amount of earth movement that took place between the two depositions. Where the movement has been great the unconformity as a rule is easy enough to make out; when the movement has been of a more gentle nature the disconformity may be scarcely detectable at all at any one exposure. In such a case only a careful tracing of the line of junction across country can show the true relationship, for then the newer beds will be seen to lie upon different beds of the underlying series as the junction is traced. All this can be anything but easy in the field and usually calls for a sound knowledge and experience of the two rock systems. One very useful guide is the strike of the beds. The strike of a bed is a line drawn at right angles to the true dip of the strata: the strike of conformable beds is said to be congruent, in other words the strikes run parallel. On the other hand unconformable beds are said to have incongruent strikes, meaning that the strikes run not parallel to each other but at an angle.

Since what is usually implied by an unconformity is the transgression of the sea across a sinking land surface, very often detritus eroded from the exposed rock finds its way into the deposits being formed, and at times even beds of pebbles, or conglomerates, marking the ancient shoreline are laid down. Since these pebbles were eroded from existing rocks and were washed into place, we can expect to find them well rounded and lying in close proximity to each other with a minimum of cementing material between them.

It was W. T. Aveline of the Survey who said, in the Geological Magazine of 1872, that ". . . the only unfaulted junction between the Borrowdale Volcanics and the Skiddaw Slates was at a place near Bootle in West Cumberland". Well, this Bootle *is* in West Cumberland and not far from its lovely coast, so our investigation is going to take us into territory which for far too long has been almost completely ignored by the ordinary visitors to Lakeland. But, to those of you willing to venture a little further afield than usual, I think I can promise, besides some interesting geology, a glimpse of a lovely bit of Cumberland. Thinking of our West Cumberland reminds me that although the exposure near Bootle was the only one of its kind known to Aveline this can hardly be accepted as the case today, so I think we will begin with an area a little nearer than Bootle, not only for the controversy that is our immediate interest, but because of the general geology as well.

Excursions:

A. West Cumberland down to Millom

COMMENCING our excursion near Ennerdale Bridge (065156) let us make our way up the hill towards Kinneyside Common. On our left is a peep of Ennerdale lake with the stepped Crag Fell coming into view. As we near the top of the hill there is a deep cut gill to the left of the road with beds of Skiddaw Slate exposed. Crossing the cattle grid make your way to the Standing Stones, a stone circle the authenticity of which I have heard questioned. Away on our right we can see Low and High Cochow with the rough road to the farm between. At our feet is a broad channel, quite waterless as a rule, leading into the really magnificent Nannycatch. Here we are in classical glacial country and before us lies the deep cut gorge carved by escaping melt-waters at the end of the Ice Age. The views from here on a clear day are really fine, stretching from the Galloway Hills across the Solway to the North to a glimpse, sometimes, of the Isle of Man in the West.

Continue along this road, which runs roughly parallel with Nannycatch, as far as Lowther Forest. If this forest is new, belonging as it does to the Forestry Commission, it is as well to remember that the ancient name of this upland area was Copeland Forest, but 50 years ago there was scarcely a tree for miles. Like many another in both Westmorland and Cumberland this old forest was cut down to provide the charcoal to smelt the iron ores which have been mined hereabouts for hundreds of years. From the northern edge of the present-day forest there is a magnificent view down Uldale towards the sea. Notice too that to the right of the forest edge, and below the farm of Lagget (055126), a flat topped area stands between the forest and the gorge. This is an outwash delta and it was laid down by debris-charged water from the melting ice at a time when the melt-waters were dammed back to form a small glacier lake.

But on now to the research that brought us here. If you follow the road by the forest for a little way you will come to an entrance to Lowther Park (061120), as the forest is called, and here there is ample room to park a car. Turn now towards the east and the hill known as Swarth Fell. If you cross this fell, just below the summit you will find yourself looking down into the valley of the river Calder. The hill directly opposite is Latter-barrow—notice that this hillside is marked by ridges of dark rock. This is the Latterbarrow Sandstone, and although the rock is almost totally hidden from view by a coating of Glacial Drift the hill on which we are standing is composed of the same rock.

If you follow the course of the stream below with your eye you will notice that a little way upstream there is a sharp, almost rectangular, bend (068118). This is where we wish to start our investigations. Down by the stream, approach the bend on the upstream side watching for outcrops of rock, and just before the stream turns down the valley again there are beds exposed. Beds of dark Skiddaw Slate are interbedded with thin bands of more silty material, in fact there are bands of a hard siliceous sandstone. A little search in the darker shale may well be worth-while for some time ago trilobites were found here, and trilobites in the Skiddaw Slates are extremely rare. One of them, a Cyclo-pygie, is now in the Museum of the Geological Survey in London.

Turn downstream now and a few yards below the bend examine the rocks carefully. You will find a completely different rock to the ones you have just been examining, for this is an igneous rock, a dyke intruded into the Slates. The form of this intrusion is perhaps best seen across the stream in the left bank but it can be found on the right bank if the stream is high. I think that one should go prepared to cross the stream, however, for a few yards further down in the left bank there are interesting exposures. While there is Skiddaw Slate exposed, the inter-digitated bands of sandstone are now much thicker than we saw before, and this is the Latterbarrow Sandstone. Unfortunately drift obscures further exposures on this bank so all the other relevant exposures are to be found on the steep banks which form part of Swarth Fell. In spite of much drift, exposures are not too difficult to find, but the rock exposed proves to be extremely variable. A tough hard siliceous sandstone that varies in colour from grey to brown is perhaps most common, but higher up the bank a little search will disclose a band of rock which can only be described as a conglomerate.

As you make your way along this bank in a downstream direction, and carefully watching for exposures, you can reflect on what we have seen so far, and what it suggests. It would seem that here is evidence of a distinct shallowing of the ancient sea in which the Skiddaw Slates were being laid down and that somewhere rock was exposed to supply the material for a conglomerate. But the streamcourse we are following is narrowing now and the banks are getting steeper. Here we must watch out with great care for every exposure, for at one a tap will show Latterbarrow Sandstone while the next, maybe only a foot or so away, shows a different rock. It is true that the base of this rock is still a siliceous sand, but intermixed with it are sharp angular fragments. Carefully examined some of these fragments are obviously Latterbarrow Sandstone, but the rest are angular bits of volcanic detritus, and the further along this fell-side we go the more prevalent these fragments become until the rock we are finding can only be described as a volcanic breccia. What seems to have happened here is that beds of sand were accumulating in a shallowing sea when from some nearby volcanic vent material was ejected to fall into the sands of the sea bed.

With great banks of drift forming one wall of the valley we come now to a narrow gorge and waterfall, Gill Force (066113), where the weather has taken advantage of the weaknesses caused by a fault or faults which throw the sandstone against tuffs. If you cross the stream above the fall and examine the bank immediately above the gorge a most interesting exposure can be seen, for here a bed of what can only be called agglomerate can be found. The exposure is only small, and the constituent material is very coarse, while the fragments are mostly of Latterbarrow Sandstone itself. Why this small isolated exposure is here I do not know for the real vent is some three-quarters of a mile away across the fell.

Continue downstream, but heading slightly uphill away from the bed of the Calder, and you will come to a tributary stream, Latterbarrow Beck. Crossing this continue over the fell until you come to the next stream; this is Caplecrag Beck, and up on the fellside beyond a rough rocky outcrop can be seen perhaps half a mile south-east from the Calder-Latterbarrow Beck junction. This exposure (072102) shows a very varied assortment of rocks, from a coarse conglomerate or agglomerate with pieces above a

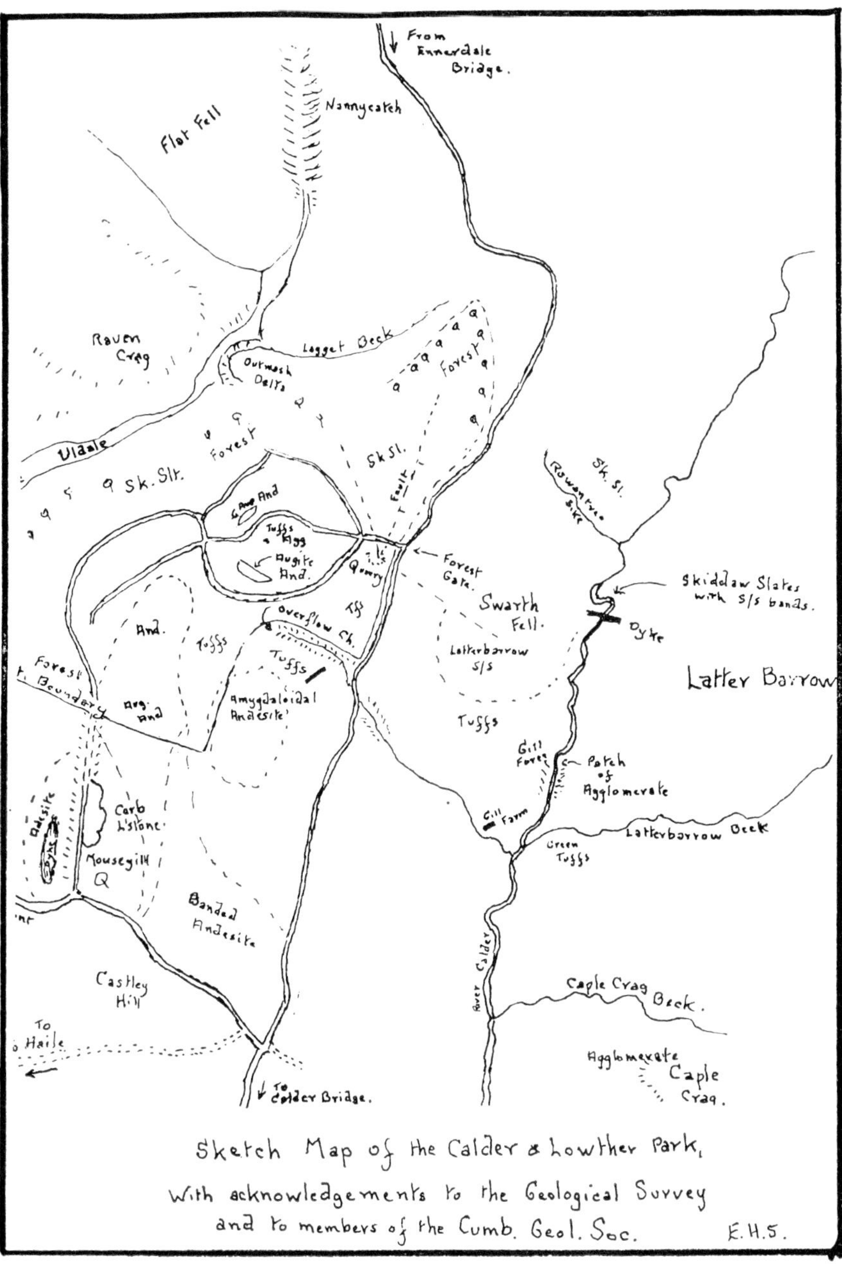

Sketch Map of the Calder & Lowther Park,
With acknowledgements to the Geological Survey
and to members of the Cumb. Geol. Soc.   E.H.S.

foot across cemented into an ash base, to much finer material interbedded with the conglomerates. While perhaps most of the larger fragments are of Latterbarrow Sandstone, some are undoubtedly andesitic, and there are occasional blocks of what seem to be a flinty rhyolite. Most of these blocks could only be described as sub-angular and they usually lie with their long axes parallel with the bedding. This is inclined to the south-west at an angle of some 50 degrees, thus giving the impression that the material has been water transported from a nearby source. A very similar deposit can be found at about 1,000 feet and some 1,200 yards further east, suggesting a continuous deposit under the peat of this upland. Here of course we are dealing with an actual vent from which issued volcanic material long, long ago, and as with any volcanic outburst a passage had to be blasted through existing rocks before the material could reach the surface. So it is not surprising that much of this exposure is made up of what was, at the time, the local rock—our Latterbarrow Sandstone. The bigger the pieces the nearer the vent they fall, so obviously if not on the actual vent of this ancient volcano we are here at least very near to it.

If on your return you watch the exposures low down in Latterbarrow Beck, and in the bed of the Calder below the junction of the two streams, true volcanic tuffs can be seen and these are followed downstream by andesitic lavas, some of which are beautifully amygdaloidal. If you now return towards the main road by the farm at Gill and make your way back towards the entrance to Lowther Park, don't fail to notice by the southern edge of the forest a marked glacial drainage channel (058115). All along the brow of this channel there are exposures which show some fine volcanic tuffs, sometimes with bits of the sandstone, and quite often wisps of reddish mud in the base of volcanic dust. Polished they make lovely specimens.

Within the forest itself there are many exposures of interest. On the left, in a hollow just before you reach the side road (060119), there is a fault exposed with Latterbarrow Sandstone on one side and tuff on the other. Further up the road there are many beds of volcanic detritus—some fine, some coarse—while still further over these basal tuffs seem to be missing but undoubted Borrowdales seem to lie directly on to the Skiddaw Slates. Although these exposures are far from simple and have in fact been argued about ever since they were first noticed by J. C. Ward, what seems of real importance to me is that in them you can see firstly Skiddaw Slate in the valley of the Calder. Then these same Skiddaw Slates are interbedded with a grit, the Latterbarrow Sandstone. But as you traverse the fellside this sandstone begins to be invaded by fragments, both of the sandstone itself and also of volcanic ejectamenta. Not far away we have seen evidence that suggests a local vent which was formed when volcanic forces burst through the existing Latterbarrow Sandstone.

In this area, then, it would seem we had a shallowing sea in which sand was being deposited above beds of mud, and these mudstones have sparingly yielded upper Skiddaw Slate graptolites like Glyptograptus. These beds of sand and mud had already consolidated when vulcanicity broke out, and the explosion scattered both broken bits of the sandstone and fragments of volcanic material to form agglomerates near the vent and volcanic tuffs over a much wider area. That later earth movements have greatly complicated this picture cannot be denied, but for the piece of research we are primarily engaged upon these same movements are largely irrelevant.

As it will now be far too late to take the road to Bootle to continue our investigations we may as well round off the visit to the district with a little more geology. Take the road south until you come to the crossroads at Cold Fell Gate; here turn right and follow this road for about half a mile to where a rather rough lane leads off to the right. From the road junction, if you look around, you will be able to see quarries not far along the lane at Mousegill (049110), and the rock exposed is the lowest of the local Carboniferous Limestones, the Seventh.

Mousegill is well worth examination on several counts. In the first place it is a very good example of the geological structure known as an "outlier", a mass of rock entirely surrounded by older rocks. To the west the limestone is faulted against the Borrowdale Volcanics which are exposed on the ridge to the left of the lane. To the east, behind the quarries, although no actual junction can be seen, the limestone overlies the Borrowdales, probably with a marked unconformity. The limestone itself is the lowest of the series and the quarry is in the lowest beds of the limestone; in fact in some small exposures before you come to the main quarries there are pebble beds, and these are basement conglomerates of the local Carboniferous. The limestone itself is interesting as it is quite fossiliferous with both corals and brachiopods, one of which, Davidsonia (Cyrtina) carbonaria —characteristic of the lowest bed—has been found here. When you have seen this interesting little area, which I present as a little bonus for the excursion, you may return by way of Wilton to reach Egremont in about two miles. On the way you will pass the famous Uldale iron mines which exploit one of the finest deposits of Haematite in the country.

*   *   *

Bootle, with its ancient church, stands on the line of the old Roman road under Black Coomb and is approached from the north by way of Gosforth and Muncaster amid some delightful country. Pass through its old world main street to where, almost at the end, a lane leaves between the houses on the left to a little corner called Newtown. This lane runs up alongside Crookley Beck to come out on the fell road over Corney, but it is difficult if not impossible by any ordinary car! The metalled part of the road finishes at a gate above which it divides, the left hand branch going to Fellgreen Farm. Here on the right hand side of the road stands a farm building and a gate (116885). Through this gate at about 50 yards distance, rough rocky outcrops can be seen, which to the practised eye are fairly obviously Borrowdales. In fact this is a very good example of the change often brought about at the surface by the underlying rocks. The rocks at this exposure in the field, according to Dr. Soper of Sheffield University who has carefully checked them, belong to what he calls "the mixed group" for above a basic andesite there are intercalations of both tuff and mudstone. Examine them by all means but I doubt if you will be able to distinguish between them by just hand specimens for, as Dr. Soper rightly says, they are not easy to interpret, at least in the field.

Passing these outcrops go straight across the field to the ravine known as the Great Cauldron (117883). The exposure you want is in the steep crag above the stream and great care is needed for I must warn you that our Lakeland streams are invariably very cold! I readily admit that this is a geologist's exposure, rather than the amateur's, and it will probably take a little time to sort out the two rock systems and find the actual junction. However, careful observation coupled with a little judicious tapping will help to distinguish between the two rocks, to find the line of junction and to see that there is no break between them.

As this location is not too far from where I live it is a place I have often visited, and I must confess that on looking round it is evident that this is geologically far from straight forward. You may remember I have said before that besides examining an exposure in close detail it is always as well to stand back and take a wider view. If you do that here I feel sure you will agree with my preceding remarks. When facing the stream, to the left of the steep path that leads down towards the cauldron, there is a wall of rock exposed; this is above the junction and is B.V.S. But look now to the other side of the path. The rocks exposed here are obviously well bedded gend so far as one can see with a lens appear to be Skiddaw Slates, admittedly with a considerable admixture of tuffaceous matter. This exposure you can now trace across the dell to the downstream side where another marked change brings you back on to volcanics. If you follow this exposure down to the stile in the wall you will find a most interesting outcrop for the Borrowdales are here in contact with well bedded

Skiddaw Slates. Obviously there must be some faulting involved here and I think it might well be worth while to try and dig out the location by the stile. Is this an actual junction? It would appear to be so. No, Dr. Soper did not mention this, and I wonder if in fact he has seen it.

Back now to the exposures described by Dr. Soper, for further upstream there is a fine rocky outcrop known as Washdub Crags (118884). This was unknown to me before the publication of Dr. Soper's most interesting paper, and while I do realise that a great deal of work must have gone into his investigation, again I am not so sure just what we less experienced people can make of it. However, at the base we are told we have a blue and buff mudstone, our Skiddaw Slates. If you break off a bit and examine it with a lens you will find what look like minute white specks in a blue grey base. Above this is about four feet or so of andesitic lava but without any very marked structure. This is overlain by some 18 inches of a fine tuff and above this there is what is described as a "mudstone conglomerate unit", the pebbles of which are apparent in places. This unit, which is about ten feet thick, is overlain by about some six feet of very hard tuff and this by a more markedly bluish tuff beneath a pyroxine andesite. Whether or not you are able to distinguish all these beds for yourself I think it is pretty clear there is here a very interesting section to be seen, but to me it seems to say more about the early phases of the Borrowdales than it does about the actual junction in which we are interested. Now had the conglomerate been situated just above the Skiddaw Slates and below the lowest andesite . . .!

As you return towards Fellgreen notice that a little way above the farm there is a really magnificent glacial overflow channel cut for the most part in solid rock. Whether this was a marginal channel as the older workers would claim, or a sub-glacial one as Dr. Smith would aver, to cut such a gorge in solid rock must have taken a very long time indeed. Back in the village head south. Before you on the left rises Black Coomb, a very extensive fell carved from Skiddaw Slates which have a roughly anticlinal structure. After surmounting the rise beyond the village to come out on to the more or less level road, keep your eye open for a small quarry up on the hill by the side of yet another streamcourse. This is Holegill Beck—notice the peculiar repetition in the name. A "gill" is the same thing as a "beck", and I can only suppose that a later people, forgetting or ignoring the old Norse "gill", tacked on their own word "beck".

Crossing Holegill Bridge, you will find a rough track running up the fell at right angles to the main road. If you pass through the gate and follow this track it will bring you to the small quarry (116864) on the fellside. In the small quarry you will find two rocks exposed and you may well be puzzled—one is Skiddaw Slate . . . but the other? Well, it was described by Alleyne Nicholson as the Black Coomb Granite, a rather close grained rock and I admit it looks little like either the Skiddaw or Eskdale variety. Somewhat puzzled myself, for I have never seen it even mentioned by any other geologist, I sent a specimen to Dr. J. A. D. Dickson of the University of Nottingham who has very kindly examined it for me in thin section. He describes it as a quartz felspar porphyrie, rather like some of the dyke rocks in and around Ennerdale. In texture hardly a true granite, it is probably a small offshoot from some far larger mass underlying the mountain.

As you make your way back towards the road notice a small tarn behind the farm. This is a "kettle hole", a place where at the end of the Ice Age a great lump of ice was left stranded to melt down in situ to form this shallow pool. There are many of these all the way down the west coast of Cumberland from about St. Bees to Millom. Having regained the road continue south until you have passed on your left the Whicham Valley road junction, Hodgson's Green and Baldmire, and this will bring you to Hellpool Bridge (139816). You will have to watch carefully for this for recent road widening activities have left the old bridge isolated from the present road behind a lay-by, a convenient place to leave the car while seeking our next location. Across the fields is a wooded hillside; this is Nicle Wood and is where we must look next.

I am afraid that here high summer adds difficulties, for then the fully leaved trees and the lush undergrowth does tend to hide some of the exposures we have come to see. Fundamentally there are two bodies of andesitic lava interbedded with the Skiddaw Slates, and Green's interpretation was that these were lava flows which had more or less buried their way into the Skiddaw mud. Of the lower of the two flows there are several exposures within the confines of the wood, and the lava, being hard and resistant, is easy enough to find. What are not so easy to find are exposures of the Skiddaw Slates. This is probably because, being much softer than the lava, it is more easily eroded and tends to form the lower ground.

However, it is worthwhile, even if you have to ignore the exposures in the wood, to walk round it and up to the little knoll behind the wood at the north-eastern corner (142818). Here you will find an exposure of andesite in close contact with the slates, and it is not too difficult to obtain a specimen of this lava welded to the slate. Close examination gives the impression, as Green suggested, that the mud has been dragged up into the interstices of the lava. The marked hollow immediately behind this exposure is caused by the more easily eroded slates which

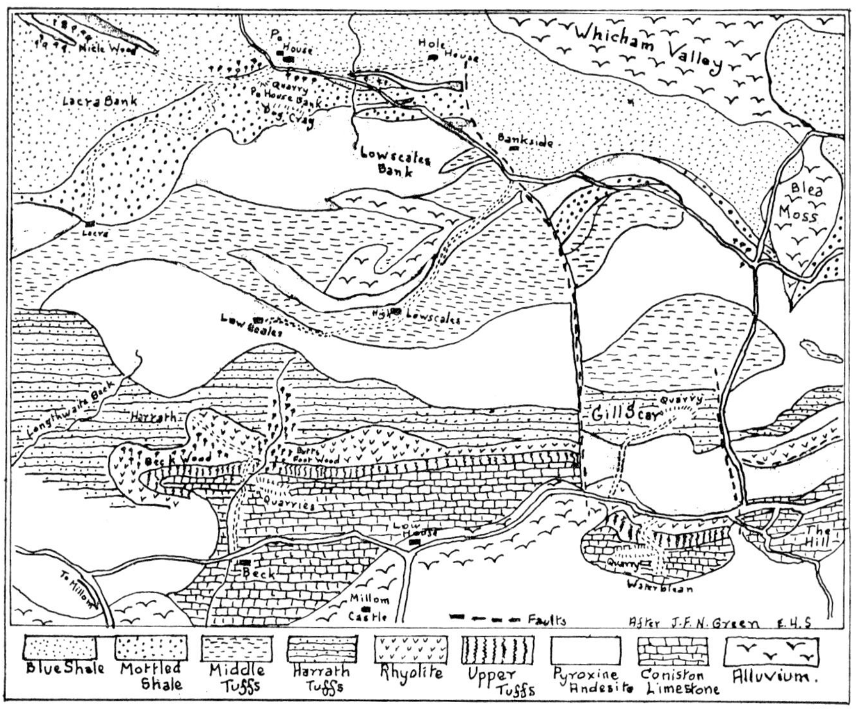

Harrath to Waterblean (after J. F. N. Green).

you can see exposed in a little gill over to the north with a steep dip roughly south-east. Up the bank and above these slates another lava exposure can be seen. To confirm J. F. N. Green's explanation would require a good deal of very careful field work, but from what can be seen on a short visit it does not seem unreasonable.

Go back now to the main road and return the way you have come as far as the Whicham road. Following this for a little way you will find a signpost which indicates "Po House". This is a rather puzzling name, I admit, for a very old site, for behind the more modern house there are older buildings that probably go back to Elizabethan times. Above the house and its grounds rises Po House Bank, and higher up still Dog Crag, both of which are of geological interest. After seeking permission at the farm go through the gate at the foot of the bank where, in a small rather overgrown quarry (149824), there are some interesting exposures. Low down on the left-hand side of the exposure there are folded and bedded rocks which on close examination are not easy to place. Either they are "Skids" with considerable volcanic detritus, or possibly they are nearer what Green calls Dog Crag Tuff. About the middle of the quarry face undoubted Skiddaw Slates are exposed but they are well brecciated as if along a line of fault. On the right there are slates overlain by tuffs but with a slickensided junction. Above the brecciated exposure a large lump of rock sticks out from the face. Examination shows this to be the blue-grey Dog Crag Tuff, but again well slickensided on the under surface.

It is exposures like this that have for so very long helped to keep this controversy alive. They are often most complicated, difficult to interpret and, to be fair, sometimes seem to invite an interpretation that depends upon your own particular views. However, the overriding consideration should always be, what evidence, if any, does the exposure provide that bears upon the *original junction*. Further up the bank towards Dog Crag there are many exposures of well bedded blue-grey tuff which, while many specimens can be found that do contain bits of "blue shale", have as a base mostly volcanic dust.

Green says: ". . . a quarter of a mile north-east of Po House a small stream crosses the road . . ." (151826). Let us go slowly along until we find this stream. Above where it passes beneath the road stands an ancient milestone which says, rather cryptically, "Millom Above" and "Millom Below", while on top of the stone there is a bench mark. Some 20 yards or so along from here you can see the "blue shale", as they for long called the Skiddaw Slates hereabouts, exposed in the bank. It is of the normal soft, dark upper Skiddaw Slate type that looks hopeful for fossils, but so far I have found none. Try your luck by all means but *don't* leave the debris lying on the road when you have finished! Moving back to the streamcourse enter the bed of the stream above the road and find, some ten feet or so above the bridge, cleaved slates exposed in the left bank. These are overlain by iron-stained beds that have rather a sandy look about them, but what is interesting is that they contain perfectly round droplets of volcanic glass, as well as what look like andesitic ejectamenta. Immediately above this, "mottled shale" is exposed, and in one or two other places in the gill beneath the fall. It passes up into the Dog Crag Tuff above.

Now it really depends upon how much time you have, but there are other interesting exposures mentioned by Green close by. "One hundred and fifty yards above the road", he says, "a well brecciated lava can be seen to have burrowed into the tuffs."

I am afraid that if you try to follow the streamcourse up for this 150 yards you will find it very hard going indeed, for all the way it is one tangle of briars and undergrowth. It is better, I think, to go a little beyond the shale exposure by the roadside to where a gate gives access to the fellside. If you gain re-entry in this way and make for the crag, in front of which there are some trees, you will find it easier but still not easy going. Beyond this first crag the ground flattens out somewhat but still further over there is a rough crag, and this is the location for the flow-brecciated andesite. The lava is well flow-brecciated, amygdaloidal and shows quartz veins and flow lines, while low down on the right-hand it can be seen to lie upon an uneven surface of Dog Crag Tuff. All along this ridge there are many exposures which show such lavas.

The two exposures we have previously seen of andesitic lava seem to me to be well worth some consideration, and where better on a fine day than here, with the Whicham Valley at our feet with its well marked sandy glacial outwash deltas, and vistas of Black and White Coomb beyond. In the quarry at Po House we were on puzzling ground with a good deal of evidence of movement within the Slates at or about the junction. There was also both brecciation and slickensiding, and while we had here fine examples there are many others of a similar kind where the soft incompetent "Skids" are overlain by the tough hard Borrowdales. In this quarry the overlying rock was a tuff, not as a rule nearly so hard and tough as an andesite, but it doesn't take much investigation to find that there is little comparison between these two rock types for "competence", as our American friends would say. But this movement is something that happened *after* the deposition of the two sediments and their induration and tells us nothing about what happened at the time of formation. Then in the streamcourse we saw, first, Skiddaw Slates somewhat cleaved. Above these was a further mudstone exposure but this time containing both bits of ejectamenta and glass droplets, more shales with ejectamenta, some of which looked suspiciously like bits of "blue shale", and these overlain by a definite volcanic tuff. This evidence, as far as it goes, seems to point to a completely normal passage from one rock system to the other without any break whatever.

Follow the road now which runs on below Lowscales Bank towards Bankside. Here on the fell above the farm there is an old quarry rather hidden from the road but containing one of our most beautiful local rocks. A rather unusual variety of the Dog Crag type Tuff, it is underlain by an andesite. The rock is rather hard and well banded, and as the bands and included fragments are stained either red or purple, while the base tends to a green or blue, it makes a most striking rock. About a mile further along this road one can turn left for Mirehouse and Lanthwaite Bridge to come out on to the main Whicham Valley road. Here turn right and, with Brocklebank Wood across the fields, come in not more than half a mile to a sharp rise up Baystone Bank. Here a farm road leads off to the left as you come up the hill and has a signpost to the farms. As it is no great distance I think it would be as well to walk, but before leaving the gateway a glance at the walls will show a surprising variety of rocks, with "mottled shale" predominant. If you ask at the farm I don't think they will mind your passing through the yard to visit the quarry beyond.

The farm itself is worthy of a second look. Over the porch on one side there is a date around 1630, but I understand from a most interesting talk I had with the farmer that this is far removed from the original. Apparently there were connections here both with the old family of le Fleming of Rydal, and at some time in the distant past with Conishead Priory. Local legend has it that it was here that the last wild boar in Cumberland was killed, and on the gable end there is a plaque let into the wall showing the figure of a boar.

However, a short walk brings us to the reservoir and on the right, just beyond the little building, is our quarry (172858). Like most such places nowadays it is becoming a rubbish dump, but on the right as we enter "blue shale" is well exposed, while on the back wall of the quarry is the most interesting exposure in the shape of a sloping slab. What it shows is well worthy of study. Seen at its best on a wet day, at the base is "blue shale", as the locals call our "Skids", with above this bands of sharp angular volcanic fragments, not unmixed with pieces of the "blue shale" itself. Here it would seem we have preserved a clear example of what John Clifton Ward was talking about so many, many years ago: "The close of the Skiddaw Slate period was marked, not by any upheaval and consequent denudation, but by a gradual coming in of volcanic action over the old sea bed." Here, plainly to be seen, is the evidence of what happened in this part of Lakeland.

Dark undisturbed mud was being laid down on the sea floor when an outburst of volcanic activity began, scattering bits of the rock through which the outburst took place, as well as sharp angular bits of volcanic ejectamenta and volcanic dust far and wide to fall into the soft still unconsolidated mud. Then followed a period of quiescence and a further outburst, this sequence of events to be recorded here for us all the way up the slab. Dr. Soper has pointed out to me that microscopic examination shows that my "blue shale" at the bottom of the slab does contain some volcanic dust. This may well be, for we shall see shortly that a constant increase of this dust completely eliminated the original mud altogether.

If you want a specimen recording this volcanic eruption of so very long ago as a memento, a little searching about the quarry bank should soon provide one for, fortunately for us, subsequent earth movements have imposed a cleavage which causes the rock to break *across* the bedding. But please, please do not hammer the slab itself! It is a pity that the bank above is somewhat drift covered, tending to make exposures few and unreliable. One can never feel quite sure whether any particular "exposure" is solid rock, or just a large half buried erratic block. On the whole from what can be seen it seems to me that "mottled shale" tends to pass upwards into Dog Crag Tuff, in which the Skiddaw Slate mud base has given way completely to volcanic dust.

What does seem clear from our explorations in this part of Cumberland is, that despite faulted junctions, such as we saw at Po House, the evidence points very strongly in favour of a perfectly conformable passage between the Skiddaw Slates and the overlying rocks of the Borrowdale Volcanics.

Excursions:

B. The Borrowdale Volcanic-Skiddaw Slate Junction: East and West

---

IN the *Geology of the Northern part of the English Lake District,* Ward mentions a gill to the south of the summit of Eycott Hill. As it was many years since last I saw this particular exposure I thought I would go back and check, for experience has long since taught me that exposures change as time goes by, more particularly exposures in a streamcourse. I think this is something that should be borne in mind when attempting to follow up exposures that someone has described: what you can see today may well differ from what some earlier worker saw, and both the exposures I am about to describe bear this statement out.

There are two possible approaches to the gill I have in mind but, having tried both, I fear there is little to choose between them for both are hard going. Taking the road from Keswick towards Penrith, pass the road to Heskett Newmarket and go on past Troutbeck. In rather less than a mile there is an opening on the left and a farm road that leads to Greenah Crag Farm

(396284). From the farm you can cross the rough fields to the location we want. Alternatively you can continue along the Penrith road until a road on the left is signposted "Berrier". Following this road watch for a lane on the left just before you reach the hamlet. This lane does eventually lead to Greenah Crag Farm, but at the point where it turns sharply in this direction you can leave the road to cross the fields. Whichever way you choose it is as well to go well shod for the ground is rough and is usually very wet and muddy.

Approaching from Greenah, the Eycott ridge rises in front of you with the hamlet of Berrier away to the east. In the bottom to the west runs a stream—Naddle Beck—with several tributaries, and it is the one which comes down from just south of Eycott summit that we want. But notice that on a lower level across the fields there is a small crag crowned with trees. This is worth visiting on two counts. First, the trees themselves have grown through the rocks splitting them in a quite remarkable way. Then, low down in front of the crag, one of our prettiest of Lakeland rocks can be found. A dark lava with prominent pink felspars and green inclusions, it makes a most striking rock. However, let us proceed to the gill (389292), which cuts a cleft between Eycott Hill itself and a lesser hill to the south. At the lower end, and in the northern bank, are the exposures Ward described. A little search will show a dark igneous rock, a blue-grey when fresh but tending to weather to a brown. This Ward described as a basalt. A little way upstream, but not so well exposed now as I seem to remember it, there is a bed of somewhat altered Skiddaw Slate some six or seven feet thick, while above this another thin basaltic flow can be found. Further up the bank the more usual Eycott type lavas come on.

Below the lower flow Ward mentioned a bed of crumpled slate in the bed of the stream, suggesting a fault, but I could not find this exposure on my visit and it is probably covered up by years of downwash. However, since then one of our local enthusiasts, Stanley Gates of the Whitehaven Technical College, has dug out this part and exposed the crumpled slates as Ward described them. Perhaps we can hope that at some time soon all these exposures might receive a clean; meanwhile enough is exposed to show that lavas are interbedded with Slates. This succession of rather basic Borrowdale Lavas, it is thought, issued from some vent to the north and runs round behind Skiddaw and the Carrock Fells as far as Binsey in the west, where again there are many interesting exposures. If you have plenty of time, however, a wander along the bed of the major stream may be worth while for I am pretty sure that, many years ago now, I saw an outcrop of Mell Fell Conglomerate in the stream bed.

While on Eycott there is another location of interest to which

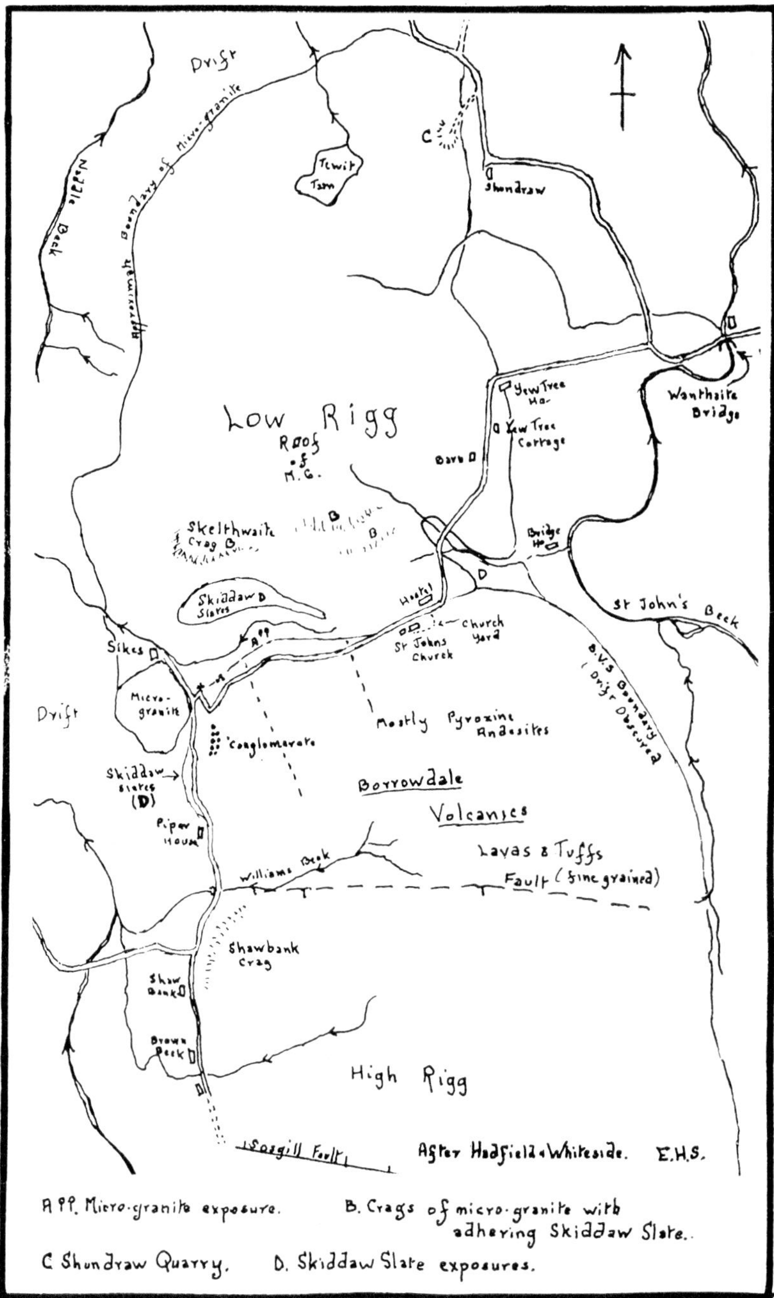
Drift
Sinen Beck
Approximate Boundary of Micro-granite
Tewit Tarn
C
Shundraw
Low Rigg
Roof of M.G.
Skelthwaite Crag B
B
B
Skiddaw Slates D
A99
Sikes
Micro-granite
Drift
Conglomerate
Skiddaw Slates (D)
Piper House
Williams Beck
Shawbank Crag
Shaw Bank
Brown Beck
Yew Tree Ho.
Yew Tree Cottage
Bave
Wanthaite Bridge
Bridge Ho.
Hostel
Church Yard
St Johns Church
St John's Beck
B.V.S Boundary, Drift Obscured
Mostly Pyroxine Andesites
Borrowdale
Volcanics
Lavas & Tuffs
Fault (fine grained)
High Rigg
Sosgill Fault
After Hadfield & Whiteside.   E.H.S.
A 99. Micro-granite exposure.
B. Crags of micro-granite with adhering Skiddaw Slate.
C Shundraw Quarry.
D. Skiddaw Slate exposures.

the late Professor Hollingworth drew attention during his re-survey of the area for the Cockermouth sheet. This exposure is in another little gill below the crag known as Fairy Knott (383305). To reach it you must follow the Berrier road north from the hamlet. Not all that long ago there were little quarries on each side of this road, but many are now either filled in or badly overgrown. Still, if you try I think you might locate places where the rock shows, and this you will find to be Carboniferous Limestone. The outcrop is part of the younger ring of rocks that entirely encircles the older rocks of the Lake District, and they lie with marked unconformity upon them. Although you can see the Eycott lavas just across the field there is no exposure of which I know that displays this unconformity.

Along this road, if you keep a sharp lookout, you will see an iron gate across a road that runs downhill on the left. Going through this gate follow the road to where there is a sharp bend, and from here strike across for Fairy Knott. The exposure is in the gill that comes down behind the Knott at a small waterfall, and this is the section as it is recorded in the memoir:

	ft.	ins.
A bluish andesite with a vesicular base and including fragments of coarse tuff	2	0
Fine grained greenish tuff	5	0
Fine tuffaceous shale	0	1
Blue-grey tuffy sandstone	3	0
Shale	0	2
Hard bluish sandstone	2	0
Greenish grey shale	0	4

A gap of some six to ten feet to grey slates below.

This, I have no doubt, is what the learned Professor saw, but that was some 40 years ago! What I found was a small fall almost entirely draped with moss, and a stream bed now encumbered by stones. I did my best patiently to clean up the exposure, and some at least of what is described above could be seen, but the lower part of the exposure is still completely hidden and, unless someone is prepared to give considerable time to it, is likely to remain so. Nevertheless, enough can be seen to suggest that sedimentation and the onset of vulcanicity were going on contemporaneously.

In making your way back to the road down which you came, notice that there is an exposure hidden from the track by a little hill. Examination will show Skiddaw Slates. As this exposure is not mentioned at all by the Surveyors I can only assume it wasn't there in their time, and I guess it is the work of local farmers looking for easily accessible road material to mend the track. I feel sure this exposure would repay a careful examination.

Our next locality is in the vicinity of the hill known as Binsey, and you can reach it by way of Heskett Newmarket and Caldbeck, or on another day from Keswick. If from Keswick follow the Carlisle road as far as the Castle Inn crossroads and turn right on the road to Uldale. The fell rising to the left is Binsey and the beds of lava of which it is made can be seen quite well from the road, but never better than when there is a thin covering of snow. This, by the way, is a point worth remembering; if it is true that our always lovely Lakeland excels when she gets a thin dressing of snow, it is also true that same snow, providing it is not too thick, often helps to show up geological structure marvellously! Following this road, then, watch out for Whitefield Cottage which stands on a road junction with a sign by the wall that says "Overwater Hotel". About 100 yards down the road to the right stands a new bungalow in what was an old derelict quarry (238349).

If you ask nicely at the house perhaps they will allow you to go behind the big building to see some really good exposures. You will find exposed at the foot of the steep bank a very fine blue-grey amygdaloidal andesite. Above this there is a thin band of slate interbedded with the lava, with still more slate exposed above. At the other corner of the quarry there are further exposures and it was from here that Green got the specimen which he had sliced and described thus: ". . . in that from Whitefield Cottage the bedding of the mud (which contains a layer half a millimetre thick, of coarser tuffy material) tends to follow the curve of the contact, and no dynamic action other than vertical pressure is to be detected." The photograph showing this junction is of a specimen I myself got from here some years ago.

From here the beds are shifted about half a mile to the northeast by the Whitefield Fault and can next be seen exposed on the hill called Latrigg, above the charming little Overwater. Go down the road until you come to a small plantation and follow the edge of it up until the wood ends. Here a rough track crosses above the wood with, close by, two small quarries and a larger one. In each a little searching will show Skiddaw Slates with tuffaceous bands, and an overlying lava a few feet away but not in contact. If you now go down to the foot of Overwater you will find that a stream leaving it runs, by way of a spillway, into a reservoir. In the steep bank of this spillway there are further interesting exposures. Probably the best approach is to go along the road towards Lowthwaite and then cross the fields in the direction of the hill, Armathwaite Head. I am afraid that, even in the best of weather, waterproof footwear is essential for the exposures are in the bank (257354) with the spillway between yourself and them. The beds exposed here are up at a high angle, probably dragged into this position by the big fault that runs nearby, but the Survey record this section:

Mudstone .... at least 6 feet
Andesite, coarse and vesicular .... some 30 feet
Mudstone with tuffy layers .... 8 feet
Lava, Eycott type, vesicular and porphyritic ..... 16 feet
Bedded tuff, fragments almost exclusively of
    Skiddaw Slate .... 8 feet

For about two hundred yards there is much drift, but beyond this there is a long section of normal lavas and tuffs.

I think I could scarcely do better than to quote Dr. J. A. D. Dickson, now of Nottingham University, who did much work here on Binsey while working for his Honours Degree at London University. In a letter written to me at the time he said: "The junction between the highly contorted Skiddaw Slates and the relatively undeformed Borrowdale Volcanics is in places definitely faulted, while in others a conformable passage, with intercalated shales in the volcanics, is apparent. A large fault separates the two formations over much of their outcrop. This fault has a sinuous course, cross-cutting most of the volcanic sequence and including the upper part of the Skiddaw Series welded to the base (and retaining normal stratigraphical relations with the lowest) of the volcanic series." There is little that one can add to this, for it is evident from what we have seen for ourselves that however much later faulting and crumpling has affected the rocks since, *at the time of deposition* Skiddaw mud continued to be laid down throughout the earlier eruptions to give us these mixed deposits.

Excursions:

C. Some Further Interesting Exposures of the Debated Junction

---

LET us return now to the more familiar Keswick area where there are other locations with interesting exposures that might help to fill in the picture. First we will go down Borrowdale to the foot of the well known Cat Gill. On the shore of the lake a little further south than where the gill enters there is a small cave with, all around it, fault breccia from the fault which runs down the eastern side of the lake. This fault brings the volcanics down against the Slates which you can see in the bed of the lake. If you come back towards the road from this exposure and bear south, look out for a marked bench at the top of the first rise. Tapping this with your hammer will show an andesitic lava flow sandwiched in between beds of the Purple Breccia.

The Purple Breccia is well exposed along the lake shore, in the nearby road cutting, and in the lower reaches of Cat Gill. Professor Marr called it a fault breccia, J. F. N. Green an explosion breccia, and while the two differed as to the source of the fragments, both did agree that *it was a breccia.* Recently, however, there have been tentative suggestions that maybe our two old warriors were mistaken, and in fact the rock is . . . a conglomerate! Admitting here, as I think I must to some puzzlement, perhaps I had better leave you to examine the rock for yourselves!

From the little area where cars are usually parked a small stream issues from under the road to run down the bank into the lake. Just below the point of issue some dark flaggy beds are exposed (assuming they are not, as is often the case, covered up by dead leaves). A careful examination will show these to be beds of a tuffaceous nature, with more than just a suspicion of Skiddaw mud in the base. Those of you who know your Cat Gill of old may remember there was a similar band below the wall which floods have washed away. Now proceed down Borrowdale to Grange and on to Hollows, but instead of turning up to the farm continue in the direction of Rosthwaite. On the right behind the wall there is a wood, Scarbrow Wood, and this is where we want to be. Nearby where there was once a gate there is now a notice which reads: "National Trust. No camping". Double back behind the wall here and follow it for about 150 yards. Here a rock breast (247171) some 20 feet or so high runs roughly parallel with the wall, with a holly tree in the middle. Across this rock there is a marked ledge and this follows pretty closely the line of junction.

Above and below the actual junction the distinction of the two rock series is more easily discerned than close to it. Above, the rock tends to a grey-green volcanic ash, and this is surmounted by a lava, while well below we have undoubted Skiddaw Slates. At the actual junction this distinction is not so easy to make as the Slates themselves tend to be hard, grey and tuffaceous. You can trace this junction right through the wood to the wall, and indeed up the fellside beyond to the exposure that has been known for a long time. This exposure, Dr. Soper thinks, I have tended to misinterpret as there is a fault. In fact I am glad to have his assurance of this as relations here always seemed a little difficult. On the other hand, though, this fault, as is often the case, is irrelevant, as what could be seen in the old days was a slab of rock down the middle of which, as if drawn by a pencil, ran a line. On each side of this line differential weathering had clearly brought out the contact between the Skiddaw Slates and the Borrowdales. Whether Dr. Soper ever saw this slab before folks with more enthusiasm than common sense wrecked and very largely removed it, I do not know. I had known it since the days of my youth and, however much subsequent faulting had moved it from its original alignment, there was little doubt about what could be seen! Since the above was written the lower part of this exposure has been dug out and carefully checked, revealing a perfect transition from mudstone to tuff. (P. J. F. Jeans, Proc. Geol. Assoc.)

Dr. Soper in his most interesting paper describes another exposure near a yew tree, due west of Hollows but well up the fell. This shows a thick development of conglomerate, with mudstone fragments up to some five centimetres across, below the junction. Now this is just what one should expect at an unconformable junction—pebble beds, or conglomerate with the pebbles derived from *existing rocks*. Thus this exposure seems to show that there were undoubted local unconformities. However, it is never wise to argue from the particular to the general so let us patiently pursue our explorations.

Leaving Keswick by the road south proceed down Low Nest Hill towards Shoulthwaite, but at the bottom of the hill where the road turns sharp left notice that on the right-hand side of the road there is some open space behind a wall (291221). You could have parked here in safety until recently when, for some obscure reason, someone blocked the entrance with big boulders! Hidden in the trees a small stream descends the hillside and it is in this streamcourse, nameless so far as I can ascertain, that we want to pursue our research. Following the stream up you will find a tree bough across the track. It is a bit awkward here but as you dodge the obstacle don't fail to notice that you are standing on an exposure of soft fissile Skiddaw Slates and that this continues, except where covered by boulders, to the foot of a small waterfall. This slate is of the Mosser-Kirkstile type, the kind that often contains graptolites; so far I have failed to find any but I still feel hopeful that one day someone will! A few steps now take you to the fall and it is at once obvious, even to a beginner, that this is a very different rock. A tap will show that it is in fact a tough hard andesitic lava, but immediately beneath it the soft shale is exposed. There is no visible break whatever and the lava lies snugly upon the undisturbed mudstone.

Turn now to the left and examine the bank. Here lava can be seen lying upon an undulating floor of the same Skiddaw Slates and, when I first uncovered this part of the exposure in the spring of 1969, there were pockets of the same "mottled shale" we saw so often in the Whicham Valley along the junction. Unfortunately some too enthusiastic collector removed these during the summer of that year, and again I can only protest! This sort of specimen is of greatest interest when it can be seen in situ; remove it and most of its value is gone! So please, please do consider carefully when you are tempted. In this particular case there was no need to disturb the exposure at all, at least along the junction, for if a specimen of mottled shale was all that was required there is usually any amount of it lying about. A little examination will show that there is a bed of mottled shale above the lava of the fall, with a series of volcanics exposed above.

This is what can be seen, and the interpretation put upon this exposure by many geologists over the years was that it showed a conformable passage from the Skiddaw Slates to the Borrowdale Volcanics, with ejectamenta falling into the slowly accumulating mud, just as Ward first visualised it so many years ago. However, a completely different interpretation is being put forward now. The lava block at the fall is a boulder in a conglomerate and, as this conglomerate lies between the soft mudstones of Skiddaw age below and the lavas of the Borrowdales above, the junction is unconformable!

Leaving this without further comment let us go and see another exposure not far away (see map page 49). Following the road south turn into the lane on the left and carefully cross the flat valley floor almost to Sikes, but stop before you begin the zig-zag ascent to the church. Here by the roadside on the right there is an exposure in the bank (302222). Examination will show that most of the pieces making up the exposure appear somewhat waterworn and lie fairly closely in contact. In fact this deposit comes much nearer to the kind of rock described as "conglomerate". This is what is immediately noticeable but a close examination shows that most of the bits and pieces are derived from the volcanics! There are occasional bits of slate too, but why not as the two rocks are exposed together over a wide area hereabouts even today. It is suggested that this deposit lies at the base of the Borrowdales and thus demonstrates an unconformity. Ward himself suggested that it might lie near the base of the volcanics, but he said nothing about unconformities.

So at two exposures, one on each side of the valley, we have seen rocks described as conglomerates, and while neither seems to me to be an ideal example of such a rock, the one we saw first seems more than doubtful. However, for the sake of argument let us accept both and then ask the simple question . . . so what? The discussion is supposed to be about the conformability, or otherwise, of the junction between the old Skiddaw

Slates and the younger Borrowdales. So what bearing, I think we are entitled to ask, can a conglomerate made up of fragments from the *Borrowdales* have upon the question? At the point in time we are discussing the volcanics as a rock series had no existence. Thus, a conglomerate composed of such pebbles might well have an interesting story to tell about the early development of that series but, as regards the question we are supposed to be discussing, the conformability of the two rock systems or otherwise, such a conglomerate is completely irrelevant. The only conglomerate that can have any real validity in this discussion is one composed of pebbles of Skiddaw Slate, since this is the rock that at a real unconformity we would visualise as consolidated, upheaved and weathered to provide the pebbles along the shoreline as they sank once more beneath the waves. Such a conglomerate, of more than strictly local extent, we have as yet to see. Let us not, however, come to any hasty conclusions but patiently look further afield.

The recrudescence of this age old controversy has been brought about by the study of cleavages, chiefly within the Skiddaw Slates, and, more particularly, cleavage directions. That the direction of a cleavage should be regarded as being at right angles to the applied pressure is reasonable enough, but the further inference that each different direction of cleavage implies a separate earth movement to produce it seems open to not a little doubt. That the application of pressure to a large and varied rock mass, regardless of what the mass itself may contain, and over a very wide area, must result in one invariable reaction seems more than doubtful. It must be remembered that cleavage is an aspect of the rock's lithology, and like all such lithological characters carries no definitive time indicators.

Trying to look objectively at this most interesting, if somewhat exasperating controversy, it would seem that what we lack most of all are definite horizons that can be established and maintained over wide areas. Neither in the Skiddaw Slates nor yet in the Borrowdales have we achieved this. Today much attention is being given to the micro-palaeontology of the Skiddaw Slates, and it may well be that, given sufficient time and patience, something positive will emerge which will help to rectify our present weakness with regard to these rocks. As regards our Borrowdales, various workers have attempted from time to time to cross-correlate beds of lava and ash with beds in other parts of the district which undoubtedly *look alike.* J. F. N. Green, having established his succession in the Duddon area, then proceeded to use his very distinctive Harrath Tuffs to establish a like succession elsewhere, from Borrowdale to Ullswater. But underlying all these attempts lies an assumption, and how far, I wonder, are we justified in making it?

The assumption we seem to be making is that because in area "A" you have for example, Harrath Tuff, then in area "B"— where undoubtedly beds with similar characteristics crop out— this same occurrence pre-supposes the *same eruption from the same vent.* In other words we had here in Lakeland in Ordovician times, one great period of vulcanicity which affected all our area equally, and more or less about the same time. Frankly I see little justification for this assumption and I notice that A. J. Wadge of the Survey has recently suggested that the Skiddaw Slates (i.e. their uppermost beds) get younger as one moves east of our central district. This seems to be borne out by the discovery in the Haweswater tunnel of specimens of *Didymograptus murchisoni,* while elsewhere all we have to show is *D. proto-bifidus,* the undoubted forerunner of *murchisoni.* I would be tempted to add that this not only occurs eastwards, but very possibly as one moves towards the fringes of our central Lakes. At Greenscoe, near Dalton-in-Furness, graptolites of the *D. bifidus* type have been found in the "blue shales" of the Brickworks quarry, and these underly the "mottled shales" and agglomerates of a nearby vent. *Bifidus* is also found in the Skiddaw Slates of the Southerfell range, and at Hazlehurst, at the foot of the Eycott ridge with its lavas, and in the Binsey area I have myself found *Glyptograptus.*

It seems probable, as often happens in cases like this, that there has been a tendency towards over-simplification, with a seeking, consciously or otherwise, for one sweeping explanation that can cover like a blanket the whole of our lovely district. But is this strictly necessary? After all we are dealing with a geosynclinal period when intense volcanic activity broke out, and the concomitant of this, surely, would be increasingly unstable conditions. No basin of deposition extends for ever, and no such basin is ever of equal depth all over. That we had such variations seems pretty evident if you consider such exposures as those of Latterbarrow, or of Great Scafell. On Latterbarrow, as we have seen, you can go through from Slates to sandstones and grit, with patches of conglomerate, while overlying these are beds with volcanic ejectamenta, which in turn become tuffs. This new interest in the early days of our Lakeland's vulcanicity may well add some interesting pages to our Lakeland story of the rocks, providing, and I must stress this, providing we seek the facts, the whole facts, and not just those which seem to fit neatly into the latest enthusiasm!

To one last location will I direct your steps before I leave this part of the story, and I must place on record thanks to A. J. Wadge of the Survey for drawing attention to it. Take the road down to Ullswater which runs from Troutbeck across Matterdale Common and notice that, just before you reach the hamlet of Matterdale End, there is a wooden gate half hidden on the right of the road, with a streamcourse across a field beyond (395237). This is Matterdale Beck, and it was to the middle part of this beck that Wadge recently led members of the Geologists' Association. Having missed this part of the excursion on the day I came to the area later and, not knowing where to look for the exposures described in the hand-out, I started here just across from this gate. Below is a rough record of what I found, along with my sketch map of the area.

On my first visit I failed to notice the somewhat folded exposures of Slates at location "A". Possibly my delight and excitement at recognising an old friend, the "mottled shale" we have seen so often before, blinded me to this first exposure. Let this be a warning: never, I must emphasise this, never wholly trust your first impressions, always go back and check, for as often as not you will come away surprised at what you have missed! On a subsequent visit it was a little chastening to find that at exposure No. 1 the rock didn't fully match up to "mottled shale" either. It has the same characteristics of angular ejectamenta, some of them pieces of Skiddaw Slate, in a fine base, but this base does not seem to contain quite so much Skiddaw mud as other examples—be this as it may it does exemplify an early explosive phase of our vulcanicity.

At exposure No. 2 the material was very badly weathered but I did manage to dig out a piece that closely resembled exposure No. 1. My exposure No. 3 was a bit puzzling, but turned out to be a volcanic tuff, while exposure No. 4 I found to be most interesting. Low down at stream level a blue shale was exposed. It was well bedded and ran up into the bank until, before it was buried under drift, it was about one foot in thickness. This was snugly overlain by a grey tuffaceous mudstone, again about one foot thick, with really fine slickensiding at the junction. Above this, beds of a pink coloured tuffy sandstone ran up the bank for at least another ten feet or so. I then realised that exposure No. 3 was simply an extension of this material.

Then followed quite a spell without any definite exposures, and while the one at No. 5 was rather small it did seem to show evidence of movement again. The stream then contracted to a gorge with the rocks well exposed on either hand. Several beds, each about a foot thick, lay upon each other, with the exposed rocks running through the gorge for several yards. But what was the rock? Borrowdale Volcanic—yes—but lava, or tuff? Here I must confess to the difficulty one often finds in attempting to sort out these rocks in the field, but if in this case the hand specimen wasn't very helpful, the exposure as a whole was a little more so. Firstly it was *well bedded,* and there were no steam holes or amygdales either, so far as could be seen, so I

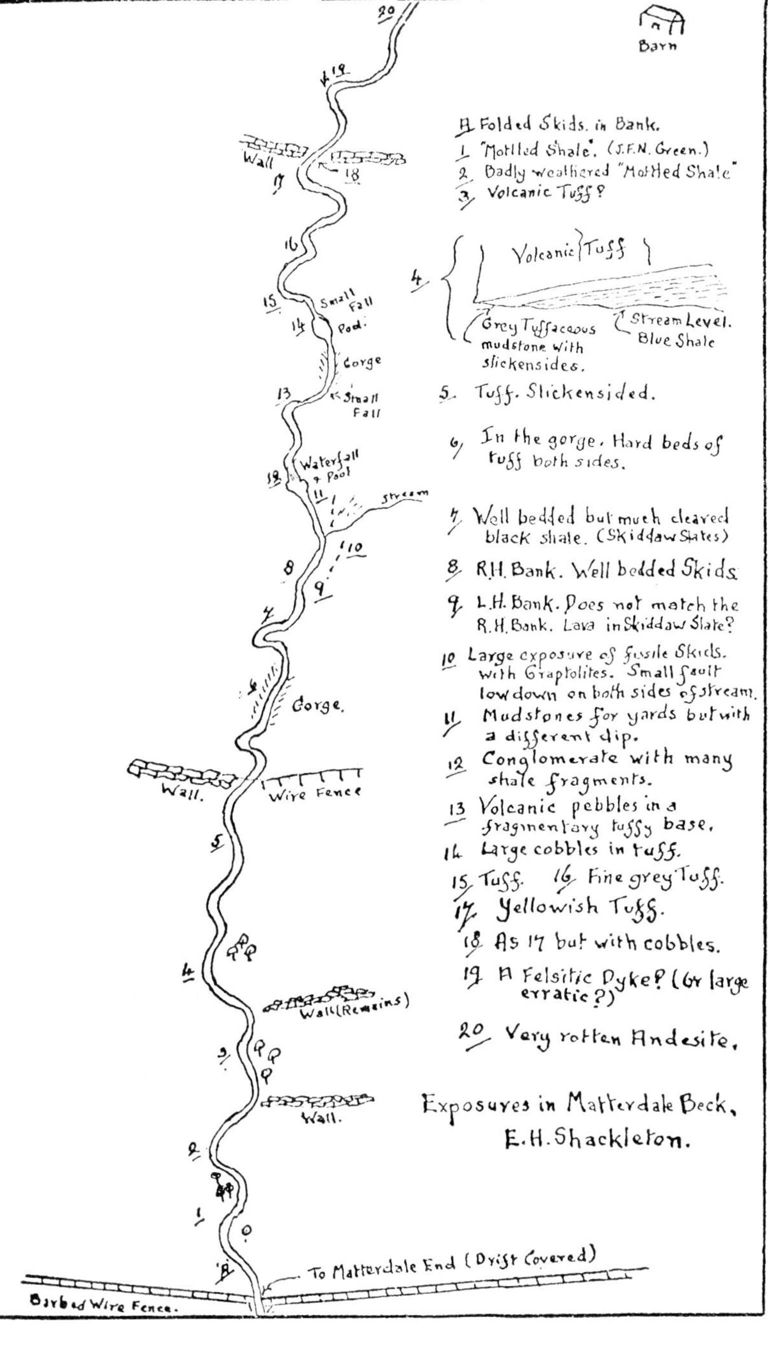

20
Barn
19
Wall
18
17
16
15 Small Fall
14 Pool
Gorge
13 Small Fall
12 Waterfall + Pool
11 stream
8
9
7
6
Gorge.
Wall. Wire Fence
5
4
Wall (Remains)
3
Wall.
2
1
0
To Matterdale End (Drift Covered)
Barbed Wire Fence.

A Folded Skids. in Bank.
1 "Mottled Shale". (J.F.N. Green.)
2 Badly weathered "Mottled Shale"
3 Volcanic Tuff?

Volcanic Tuff
4
Grey Tuffaceous mudstone with slickensides.
Stream Level. Blue Shale

5 Tuff. Slickensided.

6 In the gorge. Hard beds of tuff both sides.

7 Well bedded but much cleaved black shale. (Skiddaw Slates)

8 R.H. Bank. Well bedded Skids

9 L.H. Bank. Does not match the R.H. Bank. Lava in Skiddaw Slate?

10 Large exposure of fissile Skids. with Graptolites. Small fault low down on both sides of stream.

11 Mudstones for yards but with a different dip.

12 Conglomerate with many shale fragments.

13 Volcanic pebbles in a fragmentary tuffy base.

14 Large cobbles in tuff.

15 Tuff.   16 Fine grey Tuff.

17 Yellowish Tuff.

18 As 17 but with cobbles.

19 A Felsitic Dyke? (Or large erratic?)

20 Very rotten Andesite.

Exposures in Matterdale Beck.
E.H. Shackleton.

concluded—probably a tuff!

Emerging from the gorge and moving upstream around one of the many bends, I came to an exposure on the right bank of well bedded but much cleaved black shale. This, at exposure No. 7, I concluded was Skiddaw Slate—and why, you may ask, Skiddaw Slate? To this I can only reply that to me they looked like "Skids"—just that!—except, of course, that as man and boy I have been looking at these same wretched rocks in a great many places over a long time. Around another bend we have two exposures, No. 8 on the right bank and No. 9 almost opposite on the left. With the same presumption I dubbed the rocks on the right bank as Skiddaw Slates, dipping at a fairly high angle into the bank. But No. 9 looked intriguing. Examination showed a small exposure of igneous rock, possibly a lava, but surrounded by Skiddaw Slates.

Just a little further along the same bank a tributary stream had cut a small embayment with really fine exposures of a blue-grey papery shale (exposure No. 10) and here, if I was able to justify my "Skiddaw Slate" assumption, I had a most frustrating experience. Settling down in the warm sunshine of an autumn day I began to look for fossils, and this with some confidence, for again they *looked* the right kind of Slates in which to find fossils. Nor was I mistaken, and if the first few were poor and indeterminate, I soon had compensation for I turned out a small, but quite well preserved graptolite of the *Didymograptus proto-bifidus* type. Then, to my intense disgust, the fragile specimen just disintegrated in my fingers! However, it proved to be worth while persisting. I found more graptolites, and there were more *Didymos* and some rather poorly preserved diplograptids that looked like Glyptograptus but few, I fear, survived the journey home. However, it was nice to find them and at least confirm that we were not only dealing with undoubted Skiddaw Slates but with the higher beds of the series. Then I noticed that a small fault ran across the beds and that this seemed to give rise to a change in the dip.

At exposure No. 11 there were several yards of mudstone exposed along the left-hand wall above a deep pool, but the dip did seem different to that at the previous exposure. In the right bank (exposure No. 12) a rock which in the hand-out was described as a conglomerate was exposed. For a conglomerate the rock fragments were decidedly angular, but most of them were undoubtedly of Skiddaw Slate. The most important point how-ever was that this fragmentary rock lies *across* the upturned edges of the Slates, so here there is undoubtedly an angular unconformity. Beyond this waterfall at exposure No. 13 there were many small volcanic pebbles in a very fragmentary tuffy base. Yet another waterfall and a small gorge led to another pond and fall, with an exposure (No. 14) in the right-hand bank. This rock showed a tuffy base with many angular fragments and what could only be described as "cobbles" for some of them were almost a foot across and eliptical rather than round. When broken across the material of the cobbles was obviously volcanic, but what was puzzling was this marked disparity in size. It is difficult to imagine how any stream could carry, without sorting, anything from less than the size of a pigeon's egg to the size of a rugby football!

In the next bend exposure No. 19 was interesting but perhaps irrelevant to the succession we are trying to follow. It shows a very pretty rock, a pink felsite with many felspars. It might be an intrusive dyke, and equally possible a well buried erratic, but in any case it seems to add nothing to the subject of our research. The next and last exposure to be checked was not too helpful for the rock was badly weathered indeed. An isolated block lying across the stream was in rather better shape and, assuming it is part of the exposure, looks like a rotten volcanic. Above this, glacial drift obscures the solid rock at least as far as the road which runs down to Matterdale End.

This then is a brief description of what can be seen in Matterdale Beck and we all owe many thanks to Tony Wadge for drawing attention to it. But then comes the vexed problem of trying to explain what can be seen. The section which was shown to the Geologists' Association does show an unconformity, though the "conglomerate" it seems to me might be better described as a breccia—an explosion breccia perhaps? In any case this exposure does show the onset of vulcanicity at this time. The "cobbles" present a problem. Their so marked disparity in size would seem to rule out their being washed into place while, again, the component fragments of the enclosing rock are for the most part angular and tuffaceous. We now know, of course, from much research into present day vulcanicity that cobbles like these can be produced as a result of friction and attrition within a volcanic vent, for it would seem that "fluidization" plays a very important part in some volcanic outbursts. Would it not be intriguing to find evidence of such work here in our own Lakeland in eruptions of so very long ago?

But then we are thrown back upon the section as a whole, and I would hope that someone more expert than myself would give some time and thought to it, for here we seem to have not one junction or passage but two! I suppose the explanation that would immediately spring to mind would be—ah! a fault, or faults! But those would tend to repeat the section as a whole or at least in part. They would *not* give a totally different section! Yet this is just what we seem to have, for below our exposure No. 7 there is but little that compares with the section upstream of this. What seems to be recorded in the rocks here at Matterdale is not one onset of vulcanicity but two—and with a considerable elapse of time and quiet deposition of Skiddaw mud, with graptolites, before a second and final onset.

So, with this intriguing, if somewhat enigmatic section, I propose to leave you. If, from what you have been able to see of these junctions in and around our Lake District, you feel that by and large John Clifton Ward's assertion about such junctions still stands, despite local discrepancies, then perhaps it will fire your curiosity and release your energies to search still further into the story of Lakeland vulcanicity and its many manifestations. Maybe the time has come for a complete review of this story which, perhaps, would be better seen against the background of a mobile, active geo-synclinal belt or Island Arc, rather than a static bit of Ordovician palaeo-geography. I feel sure that two diligent workers on our Lakeland geology, Nutt and Soper, would agree with this assertion.

References

Aveline, W. T., 1869. "On the relation of the porphyry series to the Skiddaw Slates in the Lake District", Geol. Mag., Vol. 6, p 382.

Eastwood, Hollingworth and Trotter and Rose, 1968. "The geology of the country around Cockermouth and Caldbeck", Memoir of the Geological Survey of Great Britain.

Goodchild, J. G., 1885. "Observations on the stratigraphical relations of the Skiddaw Slates", Proc. Geol. Assoc., Vol. 9, pp 468-481.

Green, J. F. N., 1912. "The older Palaeozoics and the succession in the Duddon Estuary", privately published, Hayman, Christy & Lilly Ltd.

Green, J. F. N., 1915. "The structure of the eastern part of the Lake District", Proc. Geol. Assoc., Vol. 26, pp 195-226.

Green, J. F. N. "The age of the chief intrusions of the Lake District", Proc. Geol. Assoc., Vol. 28, pp 1-30.

Jackson, D. E., 1961. "Stratigraphy of the Skiddaw Group between Buttermere and Mungrisedale", Geol. Mag., Vol. 98, pp 515-528.

Jackson, D. E., 1962. "Graptolite Zones in the Skiddaw Group", Journal of Palaeontology, Vol. 36, No. 2.

Marr, J. E., 1916. "Geology of the Lake District", Cantab. Univ. Press. Recently reproduced as a photostat copy.

Mitchell, G. H., 1956. "The geological history of the Lake District", Proc. York. Geol. Soc., Vol. 30, pt 4, pp 407-463.

Moseley, F., 1960. "The succession and structure of the Borrowdale Volcanic rocks South East of Ullswater", Q. Jour. Geol. Soc., Vol. 116, pp 55-84.

Moseley, F., 1964. "The succession and structure of the Borrowdale Volcanic rocks N.E. of Ullswater", Liv. & Man. Geol. Jour., Vol. 4, pt 1, pp 127-141.

Rose, W. C. C., 1954. "The sequence and structure of the Skiddaw Slates in the Keswick/Buttermere area", Pro. Geol. Soc., Vol. 65, pp 403-405.

Shackleton, E. H., 1966-1975. "Lakeland Geology", Dalesman Books, Clapham, Lancaster.

Simpson, A., 1967. "The stratigraphy and tectonics of the Skiddaw Slates and the relationship to the overlying Borrowdale Volcanics in a part of the Lake District", Liv. & Man. Jour. of Geol., Vol. 5, pt 2, pp 391-418.

Soper, N. J., 1970. "Three critical localities on the junction of the Borrowdale Volcanics with the Skiddaw Slates in the Lake District", Pro. York. Geol. Soc., Vol. 37, pp 461-493.

Wadge, A. J., 1972. "Sections through the Skiddaw-Borrowdale unconformity in Eastern Lakeland", Pro. York. Geol. Soc., Vol. 39, pt 2, pp 179-198.

Ward, J. C., 1876. "The geology of the Northern part of the English Lake District", Memoir of the Geological Survey.

The Coniston Limestone
and its associated Volcanics

I SOMETIMES think it is unfortunate that early work on this much debated limestone was done in the Coniston area for, as a limestone rock, it is much better developed, and looks more like the rock one regards as a limestone, in the strip of country that runs parallel with the coast from Millom to Appletreeworth, near Broughton-in-Furness. It is hereabouts that its relations to other rocks can be studied, and it was here that Green tackled the problem of its succession when he had satisfied himself about the relationship between the Skiddaw Slates and the Borrowdales. From the early days of Lakeland geology there had been much doubt and debate. Alleyne Nicholson held the succession to be conformable, and, while Sedgwick seemed at first to concur, he later expressed some doubts. John Clifton Ward always regarded the junction as unconformable between the limestone and the volcanics on which it lay. I think it was Professor Marr who first suggested that the junction was faulted, and to some extent he was followed by the Survey.

Excursions:

A.   Harrath

THIS intriguing band of more or less calcareous rock can be traced right across the district from Millom near the west coast to Shap in the east. It is in the Millom area that it is thickest —as a limestone that is, for in the Shap area the series is greatly thickened by the intercalation of much very late volcanic material, while in the west this is at a minimum. In exploring the relationship of the limestone to the underlying rocks we shall have further opportunity to study the Borrowdales where Green first explored them.

If you are approaching Millom from the north a secondary road leaves the main road on the left just before you come to

Lake District scenery: Scafell viewed from Upper Eskdale (photo: Tom Parker).

Fossils, including trilobite fragments, from Skiddaw Dodd (see Chapter 1—Excursion A).

Folding and faulting on Skiddaw Dodd (Chapter 1—A).

Agglomerate on Capel Crag (Chapter 2—A).

Mottled shale on Braystone Bank (Chapter 2—A).

Exposure of Andesite on Skiddaw Slate at Whitefield Cottage (Chapter 2—B).

B.V.S./Skiddaw Slate junction at Whitefield Cottage (Chapter 2—B).

the town (161804; see sketch map on page 43). Follow this road as far as the third gate on the left, and from here you have a good view of the tree crowned hill called Harrath. Notice that on the hillside beneath the wood there are craggy exposures of a pink coloured rock. This is the volcanic rock known as rhyolite and if you are unacquainted with the type this is a good opportunity to examine it. In the chemical classification of the volcanic rocks rhyolite is regarded as acid, and this because the silica (which acts as an acid) is very high, usually around 70 per cent. Rhyolite is regarded as the extrusive form of the plutonic rock, granite.

Crossing the field to the exposure we find that in hand specimen the rock is hard and splintery, tending to break with a conchoidal fracture. Often exposures of rhyolite show marked flow banding, while flow brecciation is not unusual, but at this location neither of these features is much in evidence. The pinky colour, which is so marked here, is not uncommon but it would be a mistake to imagine that all rhyolites are pink. This is far from being the case, and if you care to examine the rhyolites around Lady Hall (192860), a few miles away, you will find the rocks there a blue-grey to almost black. They still retain something of their glassy appearance, at least when the rock is unweathered, but upon long exposure the glass tends to vitrify and in this condition can be very puzzling indeed. Low down to the left of this exposure some slickensiding can be seen, and it looks as if this small slip proved an easy path for mineralising solutions as quartz stringers are much in evidence. Over to the right there is a small quarry under the wall in volcanic tuff, but we shall see more of these later.

Returning now to the road, and making sure the gate is shut fast behind us, let us continue along the lane to Beck (163809). It may be as well to ask at the farm for I notice a new fence has been erected recently between the path and some of the ground we wish to explore. I would not anticipate any trouble as the whole hillside here is a favourite picnic area for the folks of Millom. A marked streamcourse comes down the hillside to pass under the road above the farm, and in days gone by this stream was tapped by means of a mill-race to lead the water to a mill. Using the second of two gates follow the remains of this mill-race to the point where the rough track below crosses the stream, and some fifty yards upstream from here you should find an exposure. This is a dark blue nodular limestone fairly typical of what Coniston Limestone usually looks like. It does contain fossils but here they are neither common nor yet well preserved and are mostly fragments of shells and ostracods. These latter, sometimes referred to as "water fleas", were small crustaceans with a bony carapace, usually about the size of half a pea. Beginning in the Cambrian, they are here in the Upper Ordovician becoming much more common—so much so that in the next succeeding series of rocks, the Ashgillian, there is a thin bed near the top that is thick with them.

Follow the stream round the next bend to another patch of blue limey shale. Here remains are a little more plentiful if still somewhat fragmentary, and seem for the most part to be the remains of orthids. Still following the stream there are several more exposures as far as the bridge, depending to some extent on the height of the water in the beck. From the old packhorse bridge you can see a quarry (163812) over to the right, with beds of almost crystalline limestone some 30 to 35 feet thick overlying the same limestone with shales we saw in the beck. Here the beds are almost vertical. Again, although I have searched diligently, fossils seem few and far between, and this applies equally to the much bigger quarry further over. Examination of this bigger quarry is often difficult because of flooding, but enough can be seen to show that this rock is very much a limestone!

Leaving the quarry make your way round the edge, and some fifteen paces towards Butts Foot Wood you will find a small rocky ridge. This exposure is about six feet thick and is of a distinctly gritty nature. Green called it a volcanic sand and reported fossil fragments in it. I have had no such luck, but a little further over towards the wood there is yet another exposure which looks like a fine bedded grit. As this tends to weather to a "gingerbread rock" I felt more hopeful, and sure enough I found both orthids and fragments of coral. The gritty base seems to be of volcanic material and suggests that somewhere, at the time this bed was formed, the Borrowdales were exposed.

Cross now to the wood and squeezing through the stile find, some six paces in, a bed of hard tuffs of a greeny colour—these are J. F. N. Green's "Upper Tuffs". If you examine this rock with care you will find it to contain small pebbles, which again suggests exposed rocks to act as a source. Across a deep-cut little stream course some ten feet of grey-green ash is exposed; it is hard and tough and shows but little signs of bedding. The ash is underlain by a more shaley band, rather purplish in colour and perhaps twelve to fifteen feet in thickness. It is of interest because of the many small pebbles of andesite it contains. This purple band is in turn underlain by more greeny ash beds, while in a small quarry-like exposure purple coloured ash shows both many visible fragments and rounded pebbles. All this, of course, very strongly suggests contemporaneous erosion.

Go back a little way now to join the rough path up through the wood. The path steepens towards the top and you have to climb a small escarpment of the same pinky rhyolite we saw on the hillside. Out in the field two small ridges run across in the direction of a prominent hill. The hill is Harrath and the ridges are due to two bands of rhyolite. Behind Harrath to the north there is a marked depression with, on the skyline, a very prominent ridge. This depression is caused by bands of less resistant tuffs, Middle and Harrath, and these overlie a thick band of andesitic lava, the cause of the upstanding ridge.

Cross now to the hill and a little searching will show you exposures of the rock to which Green gave the name "Harrath Tuff". These tuffs are at least one hundred feet thick beds of volcanic detritus, some of which are in no way striking, while the rest are very much so. Green's Harrath Tuff has a green slate base, with a fairly good cleavage, in which there are many coarse fragments, but its most striking feature is the occurrence of dark green to almost black patches of all shapes and sizes up to several centimetres across. These patches Green described as "variolite" and seem in hand specimen to have a somewhat fibrous structure. Originally they were glassy, and probably they are felspars which have devitrified. While the upper bands of the Harrath Tuff are free from these dark patches, they often contain white or pinky felsite pebbles which stand out well on a weathered surface. Apart from these pebbles there seems little to distinguish these more normal bands of Harrath Tuff from the Middle Tuffs below, and both have been quarried for slates in the past.

Green regarded his Harrath Tuff as one of his best, most persistent and easily recognised horizons in the Borrowdale Volcanic sequence. How justified he was in correlating rocks of other, even remote areas, that are undoubtedly similar in appearance with the Tuffs of Harrath I do not know, nor do I see how one can actually prove or disprove such a correlation. By this I mean that it would be difficult to demonstrate that all the tuffs between here and Ullswater, that have the undoubted Harrath look, were actually thrown out of the same vent at around the same time as the tuffs on this hillside here! However this may be, a glance at the map on page 43 will show that in this area one could, within the compass of a day's walk, easily see the complete local succession, from the "blue shale" at the foot of Po House bank to the base of the Coniston Limestone we have just been examining. It may be pointed out that in this part of the country most of the prominent ridges are due to exposures of andesitic lava, and should you be in any doubt as to what a pyroxine andesite looks like you need only cross the depression behind Harrath and check the rocks on the skyline to the north.

We have now seen the Harrath Tuffs, of which there are about one hundred feet; the Rhyolites that are something like sixty feet thick; and some forty-five to fifty feet of Upper Tuffs. In this succession it is the Upper Tuffs which are of greatest interest, for their development seems to be somewhat local and they tend to disappear as one approaches the Duddon. The question is: Were these Upper Tuffs merely local, or were they laid down more generally, only at a later date to be eroded away? But the answer to this question is by no means a simple one! Before leaving this hillside a look at your map will show that across the stream from the quarry a syncline, or basin, has been mapped. This poses a challenge in field work—can you trace out the rocks on the ground? It is worth a try.

Our next location is at Waterblean, for a strong fault in the Park Plantation throws the limestone down towards the shore. At Low House we rejoin the main Millom to Broughton road and in about half a mile from this point new road works have exposed our limestone by the roadside. The rock is worth examining for I have found very good fossils here, including a splendid coral. Continue past the section of the old road now used as a lay-by until you come to a wood, up the far side of which runs a road. This leads, through a gate, to quarries on Gill Scar (172829) which are still being worked, and here you can see really fresh material from the Harrath Tuffs—in the lower quarry the more ordinary looking upper part of the tuffs, but in the quarry above fine exposures of the tuff with "variolite".

If you do not wish to visit Gill Scar continue up the main road towards The Hill, but before reaching the hamlet notice an ornate ironwork gate on the right across the road to Waterblean. Walking along this road you will come to a cottage on the right, opposite which there is a small exposure. This shows, towards the shore, Coniston Limestone up at a high angle with Upper Tuff beneath it. A little way back, behind the cottage, are old quarries (176824) from which many, many tons of haematite were worked long ago. This is the only location, so far as I know, where the Coniston Limestone has yielded haematite.

Crossing now behind the cottage and the quarry you can see in several places the junction, but between the tuff and the limestone there is a thin band of calcareous volcanic sand. It is from this little band that many fossils have been recovered in the past, the coral Favosites being the most common. That wonderful old wanderer, John Bolton, writing over one hundred years ago, tells of finding corals here in the pink calcite, but I have never had any such luck. Now at Beck, which is not all that far away, we found the Coniston Limestone to be underlain by many feet of volcanic sand. There it was a quartzitic grit with bits of felspar and pink rhyolite, and what looked like volcanic detritus with odd fossils. Here at Waterblean only a little of this grit is left before we are down on to the hard Upper

Tuffs. It would seem, therefore, that most of the volcanic sand has been swept away and it is interesting to note that some of it can be seen in the lowest layers of the limestone. The Coniston Limestone itself is said to be completely unfossiliferous here. It is obviously cut by faults along which some haematisation can still be seen.

Excursions:

B.   Graystone House

RETURNING now to the main road, continue by The Hill and The Green to re-join the Whicham road at Hall-thwaites. After going up the rise you come to an entirely new stretch of road, near the other end of which a house stands by the old road in a marked dip on the right, but it is on the left that two minor roads leave (187869). One goes to Ash House but it is the other one we want for this rather rough road leads beyond the gate to Graystone House. If you ask nicely at the house they should give you permission to go through the yard to the fields beyond. Leaving the yard by the five-barred gate follow the rough cart track for about one hundred yards to find a small exposure by the side of the road. I suggest you should examine this exposure with care, for while the rock is a rhyolite, I fear it bears but little resemblance to the lovely pink variety we saw around Harrath. This rock has been heavily weathered and de-vitrified with the results you see. If you can get a fresh break it has a slightly glassy look, but it does *not* at first sight suggest a rhyolite. Pass on to find, in another 25 paces or so, another exposure. This too is rhyolite and shows a tendency to weather white, which is more typical of rhyolites, and while this does look a little more like it, I can only say it is far from what the text books describe. This, if you will forgive my saying so, is the kind of thing where no amount of reading can help, and "go and see" is the only road to expertness!

Two more gates and we pass into a field with a big ash tree, and on either side of the track more or less devitrified rhyolites are exposed. Following this track across the fields for about 200 yards brings an old lime-kiln into view on the right, and a little knoll on the left. This lime-kiln is interesting for over the greater part of its exposure Coniston Limestone is worse than useless for burning. In most places the mud content is so high that the product of burning is a slag, rather than burnt lime fit for the fields. One of my local friends tells me that long ago there was a cement works down at Broughton Mills in the valley of the Lickle. This was of course powered by a water wheel and the cement was a product of burnt Coniston Limestone. So perhaps it would be as well to find out for a start how the limestone here differs from its neighbours.

Below the lime-kiln there is a marked trough which runs in a north-easterly direction for about 150 yards, and along this the limestone has been quarried. On either side of the trench there is still limestone left and a tap with your hammer will show a beautiful white crystalline rock which is almost marble like. It was first discovered and described by Adam Sedgwick in 1822 and named by him the "Graystone House Limestone", for then its age and affinities had not been worked out. Bolton referred to it (Bolton J., 1869, "Geological Fragments — from rambles amongst the rocks of Furness and Cartmel"), and says he found fossils here over one hundred years ago. It was about this time that it was first mapped by the Survey. Half way along the trench there is a little knoll, above a water hole, and here the limestone has become de-calcified and more resembles the rock usually associated with its name. As for the fossils I can only admit defeat, for although I have made many a visit over the years and searched carefully I have never had any luck. Of course one need not be too surprised at this, for with quarries that have lain unworked for ages all the available fossils have long since been collected!

But now let us turn to what brought us here: the limestone and its field relations. If we climb the north-western lip of the trench near the lime-kiln we shall find a ledge some two or three feet wide. Tapped on the quarry side you will find the limestone, but a tap against the bank shows a band, variable but mostly thin, of a quite different rock. It looks like a pale grey grit, but with many pink fragments visible, and this can be traced along the bank and under the trees. In some places you can still find traces of a pink calcitic rock with fragments, and I can only suppose that this is the remains of the band from which fossils have come in the past. It would seem that this thin band, or at least what is left of it, is all that remains at this location of the many feet of volcanic sand we first saw underlying the limestone at Beck.

Now let us work backwards towards the farm on the hillside known as "Boadhole", and try to check on the succession. The first obvious exposure is at the little knoll we first noticed when we came here, and lying only a few feet behind the line of the exposures we have just checked. Examination shows this to be weathered rhyolite, so the thick development of Upper Tuffs we saw in Butts Wood has completely disappeared. The volcanic sands seem to have dwindled almost to non-existence, to bring the limestone down practically on to the rhyolite. So it would seem that, either our tuffs and thick volcanic sands were never deposited here-abouts, or they were laid down but were swept away before the limestone itself was deposited. The exposures we have checked so far relating to this interesting problem are certainly suggestive of an unconformable junction, but perhaps it would be wise to follow our junction a little further before we decide.

Meanwhile, if you are interested, the farm of Boadhole itself stands on an exposure of Harrath Tuff while the ridge behind the buildings is andesitic lava in its upper part. So it would seem that the succession here is not quite the same as the one we examined at Harrath. The tuffs seem to swing round for they can be found again in exposures beneath the trees in the little beck. Apart from this swinging round out of the line of strike, the tuffs of Boadhole seem to extend towards the limestone no further than the wall in front of the house and, while no junction is exposed, the rest of the outcrops from the wall to the little knoll are of rhyolite.

To this there is an interesting but rather puzzling exception, for towards the Stanley Wood end of the limestone exposure the under-lying rock is andesite, not rhyolite. This may well be an indication of the complications introduced here-abouts by the famous Whicham Valley Fault which has caused debate for so long. So far as I know there is nowhere in this locality where the actual fault can be seen, for the ground is heavily drift covered, yet on the bank behind the lime-kiln you will find the Borrowdales exposed! This, of course, is all wrong, for the lime-stone should be overlain by the Ashgill Beds, admittedly with a subsidiary volcanic ash band. The exposures up the bank, how-ever, are certainly not of this ash band for the volcanic beds continue for a considerable distance towards the sea. This implies a strong fault with a big downthrow to the south-east. While the actual fault seems nowhere to be visible, there is a well marked trough in which stands Ash House and, curiously enough, built into the walls of the old lime-kiln there are many blocks of a fault breccia.

From Graystone House eastwards the exposures of our lime-stone are either very poor or non-existent until we come to the next area of interest near Broughton-in-Furness. Back on the road and heading for Duddon Bridge you will find some really fine exposures of the volcanics at the top of the hill which have been revealed by recent road widening operations, while at the bottom of the hill there is a roadside quarry on the left well hidden by the woods. Here a fine volcanic tuff is exposed, and these exposures clearly demonstrate the strong displacement of the Borrowdales.

Excursions:
C.   The Coniston Limestone of Appletreeworth

TO those of my readers who still have an interest in our limestone and its associated volcanics I now propose a venture into a part of Lakeland little known to the ordinary visitor, but a part which is nevertheless as lovely as it is geologically interesting. Crossing the old county boundary at Duddon Bridge, follow the road along the flat and up the hill to High Cross. Here at the four road ends turn left to Torver and Coniston, but after passing the park of Broughton Towers on the right take the first road on the left, sign-posted "Broughton Mills". A little way along this road brings you to a sharp left turn (219894) with a lesser road going almost straight on. The left turn leads to Broughton Mills but it is the subsidiary road we want. This lane, for it is little more, runs parallel with the river Lickle on the left and the Coniston Valley on your right. Of both these valleys there are fine views, while behind you the estuary of the Duddon lies spread out like a map. About $1\frac{1}{2}$ miles along this road there is a cross roads. To the left a charming lane leads down to Broughton Mills by Hobkin Ground, while the road to the right continues past the famous Broughton Moor Slate Quarries to Torver and to Coniston. However, it is the road before us that we want, and this goes down the hill to Hawk Bridge and Appletreeworth (339919).

There is a good parking place here by the bridge from where we can pursue our investigations. The hill on the left is the Hawk, part of Appletreeworth Forest, and is covered here with young red oaks which give a pleasant contrast to the conifers both in spring and in autumn when they make a marvellous show. On top of the Hawk part of the area has been left unplanted because there are the remains of a pre-historic settlement. The road forward leads to Stephenson Ground and on following this up the steep hill you will find a gate on the right which leads into the forest. Along this road there are many interesting exposures and of these I can only speak from what can be seen in hand specimen. Some 130 paces from the gate there is an exposure up the bank on the right hand side of the road; this is a slightly amygdaloidal andesite. For some little way now there are no exposures, but then there are many in quick succession and mostly still on the right hand side. These are of augite andesite but the last exposure before the wall is interesting for a close examination will show a brecciated lava overlain on an irregular surface by a tuff.

On the right-hand bend a tuff is exposed which has the Harrath look about it, and these exposures continue for a little way, but just before a new road leads off to the left we are back on augite andesite with some quartz and epidote veining. Through the trees we can see an old wall on the right as we go down to where the road swings left (with a grass track going straight on). As we go down this road we cannot fail to notice the many exposures of glacial drift in the bank to the left, but for some 150 yards there is no rock exposed. The first exposure shows what looks very much like a flow breccia, but the next, and much bigger exposure, shows flow structure, vesicles and amygdales with much iron pyrites. On the right now a road goes downhill to join the road that follows the beck by the ruins of Appletreeworth farm, while in front of us is the first exposure of the limestone we came to see. Here it shows a steep dip towards the valley, has a nodular appearance and is well worth our attention. In the first place it is fossiliferous and the first bedding plane when I first saw it was covered with fossil shells. If many of these have now, unfortunately, been removed, a little search should show both shells and corals. The exposure as a whole displays that tendency we have noticed before to weather into holes and bands of "gingerbread rock".

At this, our first contact with the limestone, exploration up the bank discloses some interesting, if somewhat enigmatic, exposures. The first exposure is some fifteen feet up the bank and many feet have trodden a path. Here the limestone is seen to lie upon a lava surface with apparently complete conformity. Now move some ten paces or so to the left and up a little, and here the limestone is seen to lie upon a lava surface that is very uneven. Is this contemporaneous erosion, or simply the original, and quite usually, uneven surface of a lava flow? However there is a further point of interest—and importance. A careful check of the dips of the two rocks, limestone and lava, does seem to indicate that they are incongruous, and this would favour erosion. The dip of the limestone is fairly easily checked but the lava will need a little further exploration. I append a photograph of the second exposure so you should recognise it when you see it ... but please, please do remember, that this does depend on the exposure *not* being unduly hammered!

Leaving this our first exposure, follow the road to another small limestone exposure which is followed by a rather puzzling one—the rock is tough and almost flint-like and I think it must be the volcanic ash band that normally lies above the limestone. The reason for this exposure may very well be faulting, for all along the valley the exposures are cut by dip faults which cause the beds to step out of line. Another small exposure is followed by a bend in the road and a slight rise, at the top of which there is a massive bed of blue limestone, again with many fossils among which I have found Halysites, the chain coral. Next we come to a large exposure, and I fear rather liable to alteration, for the quarry has been opened by the forestry people for road material. It is a very intriguing exposure and I can only record what I myself have seen. On the wall of the quarry as you first approach, well bedded limestone can be seen (and again I have found fossils), and although the bedding becomes less pronounced it does persist for some way along the wall. Then there is a sharp change—and the rock exposed is a rhyolite! This exposure will need, and certainly warrants, careful investigation. When last I was there actual junctions could be found and it was possible to collect a specimen showing limestone on rhyolite with, quite often, pebbles of rhyolite cemented together in a limy base One specimen I treasure shows a pocket of our old friend—volcanic sand!

When the quarry was first opened I saw fossil corals within inches of the actual junction and only last year fossils were found by members of a joint meeting of the Cumberland and Liverpool Geological Societies. Some of the rhyolite blocks lying about then did show flow banding and some fine flow brecciation, but what was not easy to make out was any sort of dip in the volcanic rock. However, at the other end of the quarry badly weathered limestone was exposed with a dip which seemed to coincide with the dip of the other limestone in the quarry, and it does not seem to be in the line with the rhyolite upon which it lies.

The road now makes a sharp bend and crosses a stream, just beyond which is a junction around a grass triangle with one road going down towards the beck. A little further along there is a lay-by with our limestone exposed and rhyolite beneath, but no actual junction visible. Half way towards the next bend there is a small exposure showing shale which is a little puzzling; it looks to me rather like Ashgillian, and if it is we must again have crossed one of the many dip faults. Our next exposure shows our Coniston Limestone on the left, but not much of it, while on the right of the road an old wall runs down into the wood. If we go into the wood we can find exposures some way in, and these are still in the limestone, through which the road hereabouts has been cut. Next comes a small depression of a few yards; it is caused by the volcanic ash band that overlies the limestone but careful search is needful to get a specimen. Then we come to a well marked rocky knoll, created by the exposure of the Mucronatus Band in which there are many trilobite remains, mostly fragmentary with heads, tails, eyes, bits of plurae and the occasional bigger and more complete specimen. On the brow above the stream Ashgill Shales crop out, packed with fossil shells and the occasional trilobite, while across the stream Skelgill beds are exposed on the other side of the great strike fault which runs down the valley. In fact it is this fault which is largely responsible for the valley, the stream having found and followed the line of weakness. In this exposure we have, of course, a perfect repeat of the exposures at Ashgill itself but if you do visit the place please remember that others beside yourself would like to see and perhaps collect, the odd specimen. At the risk of being thought tiresome I cannot refrain from stressing this for it is becoming every year of increasing importance.

As we make our way back towards the triangular road junction we can reflect upon all we have seen at the many exposures visited between Beck Foot Wood and Appletreeworth. On the whole the evidence we have seen can only be regarded as very much in favour of an unconformable junction between the Coniston Limestone and the underlying volcanics. On the other hand I think we must also have demonstrated how very difficult it can sometimes be to settle such issues in the field, and perhaps better understand why the arguments have persisted for so long, even among the experts. If you still wish to pursue the matter I might mention that at the well known beauty spot of Tarn Hows the limestone is well exposed on the brow to the east with the volcanic sand nearer to the tarn.

Let us now go down the road towards the beck. If we keep a close watch on the exposures on our left we shall find, about half way down the road and directly below a small larch tree, an exposure which shows a fine slickensided surface. It is just one of the many faults, great and small, with which this interesting area abounds. Turning to the right at the bottom of the descent, follow the stream in the direction of our starting point, Hawk Bridge. But do not imagine we have exhausted interest for this is far from being the case. As you go along, after some 300 paces or so, you can look out for what looks like a small quarry face up among the trees on the right. In fact this is an old adit, and if you care to examine the bits lying around you may find specimens of haematite as well as pieces of white barytes carrying specks of copper pyrites and odd smears of malachite. Having got what specimens you want (and please don't carry chunks down on to the road and leave them there for someone else to move, as I have seen on occasions!), carry on downstream past the ruins of Appletreeworth Farm.

The next exposures are very complicated indeed and I can only indicate to you what can be seen. Explanations will have to await the outcome of detailed work at present in hand by officers of the Geological Survey under W. C. C. Rose. The first rock exposed is a pink coloured shale and, when the forestry first made these exposures in search of road metal, I and other geologists thought this was the volcanic ash band that normally overlies the Coniston Limestone. However, as the exposures grew in size and in number, fossils were found, and here at this first exposure I have myself found trilobite fragments. Perhaps a cross-correlation with the Mucronatus Beds we saw upstream might be justified, although lithologically the two rocks do *not* look alike. This is yet another example, perhaps, of how unreliable determinations made on lithology alone can be!

Further along the bank the same pink shales persist and you can look out for small but rather nice specimens of dendritic manganese. However, if you watch the exposures near the top of the bank carefully there is evidence of faulting, and beyond this point many fossils have been found. These are not trilobites but graptolites; although the specimens found are on the whole but poorly preserved, enough can be seen to say unhesitatingly —Skelgill Beds. But now the stream makes a sudden right angled bend and at right angles to the pinky beds we have a bank some forty or fifty feet high. There can be little doubt as to the rock exposed here—it is the black mudstone of the Skelgills and poorly preserved fossils can be found. But what an exposure! I cannot recall ever having seen such a complicated example of multiple faulting anywhere before. A careful examination of the face will show that it is cut by one fault after another until working out anything like a detailed succession is only a remote possibility. Before this rock face was cleared (and the clearing was done at the suggestion of amateur geologists with the kind co-operation of the Forestry lads) our friends Bill Grieve and Harry Kellet had found a "green streak" exposed high up the bank. This, I confess, had me puzzled for at the type location, Skelgill, the green streaks occur near the bottom of the bank, not at the top. However, with the clearing of outcrop we can at least understand why the succession was so topsy turvy.

About the middle of the face an old adit has been exposed, and this was not visible before. Examination showed nothing in the way of mineralisation, and I was very puzzled as to why the adit had been made. Careful examination will show it has been cut along the line of one of the bigger of the many faults and there is a thick band of black fault gouge; it was only when I discussed this interesting exposure one day with Mr. Shaw of the Force Crag Mine that I got a hint. My friend recalled that when he was a lad in Coniston he had heard his elders in the village talk about a "blacking factory" which once existed—and the source of the "blacking" was these self same Skelgills above Mealy Gill. But we still have not exhausted interest here. When you have finished your examination of the shattered Skelgills, their faults, their fossils, and the not uncommon marcasite nodules, look to the left of this rock face for here the dark Skelgills are yet again faulted, and this time against a different rock. Examination may well produce a little puzzlement; if it does you can reflect you are in good company for before the section was cleared this rock had been variously described, but usually as a "pebble bed" subjacent to the limestone. In fact it is a lava, somewhat flow-brecciated — hence the suggestion of a pebble bed—and can only be accounted for by this complex system of faults.

To the left there is still another one with a thin band of fault gouge, and the succeeding rock—the Coniston Limestone! The first few beds are very badly weathered and somewhat sandy, but along the wall quite good limestone is exposed. Such a complicated section it would be difficult to match even in our Lake District, but this is not the whole story, for the rocks in the bed of the stream where it turns down towards Hawk Bridge are very suggestive of Browgill Beds. These are most certainly overlain by Brathay Flags exposed in the left bank of the stream about half way down the plantation, complete with well preserved graptolites of the *Mongraptus priodon* type (in the solid) along with *Cyrtograptus,* probably *murchisoni.*

Now at the left hand edge of our excavated exposure we have undoubted Coniston Limestone. In the ordinary succession this should be overlain by a band of volcanic ash some fifteen feet thick, then by the Mucronatus Beds, another fifteen feet or so, and these by the Ashgill Shales at least fifty feet thick. The Ashgills are overlain by the Stockdale Shales, comprising Skelgills and Browgills, together some 150 feet of strata. Now it must be obvious that between our limestone exposure and the left bank of the stream where it turns down, with the Brathay Flags exposed, there just is not room to accommodate this succession —even if the beds were up on end!

I think you will agree that in following these beds across from Millom to Appletreeworth we have seen both some very lovely country and some interesting geology, even if all the questions raised do not admit of easy answers.

Excursions:

D.   Greenscoe Crags

MY last location may not appeal to all as it is rather on the outskirts of the Lake District, but on the other hand it might well be worked in on the way home. If, back in Broughton, you take the rather twisting road south (the A595) and at Grise-beck ignore the obvious way over Gawthwaite, you can press on until you come to Askham, that centre of intense mining activities of half a century ago. After making the sharp turn left to leave the village, you can hardly fail to notice on your right the many deep pools of water, the "sops" left by the collapse of underground workings for haematite. In about a mile you will have reached our objective, Greenscoe Crags. On the right of the road is the brickworks, to the management of which applications to visit the Crags should be made, for the bricks have for many years now been made from the "blue" and "mottled" shale. Their workers first used the terms, and it is in their quarries that our fascinating if somewhat enigmatic exposures are to be found.

The whole area is full of interest for up on the crags on the left of the main road there are great quarries (now disused) carved from an ancient volcanic vent, perhaps one of the later Ordovician outbursts. There is a car park to the right of the main road; leaving your car here, follow the winding road up to the quarry as shown on the sketch map. Near where it makes its first sharp turn lava is exposed, and just beyond the next turn there is an exposure of the rock we have followed for so long, Coniston Limestone. Make your way up into the quarry. Lying about on the floor the last time I was here there were great blocks of lava, some with occluded Skiddaw Slate, and it is easy enough to get a specimen of the two rocks welded together. But in some of these blocks the occluded material is not so much obvious Skiddaw Slate as a mass of broken rock fragments—in fact a volcanic agglomerate.

If you now examine the north face of the quarry you will find that it is this agglomerate which has been quarried for road material. It should remind you of Capel Crag where we last saw a similar rock, but here some of the ejectamenta are huge, at least a yard across. Some of these fragments are, as close inspection will show, bleached and hardened Skiddaw Slate, while perhaps the majority are lumps of lava, and they were all blasted

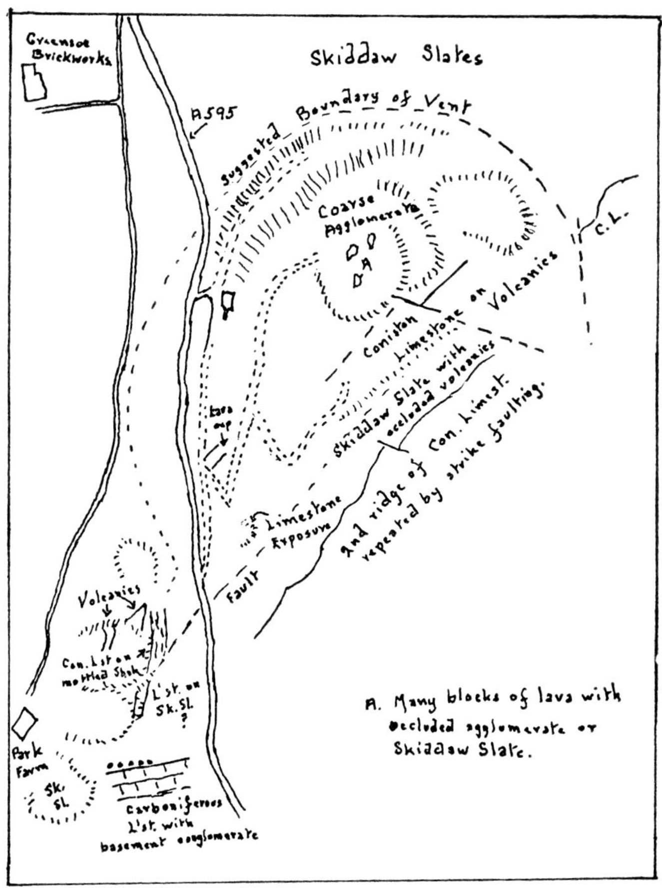

Rough Sketch map of the Greenscoe Vent and the associated Coniston Limestone outcrops

out of the vent when the volcano was being formed. In some parts of the quarry there are stringers of lava running through the agglomerate, perhaps the remains of lava flows which were on their way to the surface when the vent was last active several hundred million years ago!

When you have examined this fascinating quarry make your way out and up to the surface above. Here the rock exposed by the edge of the quarry is Coniston Limestone, and it can be seen to be lying upon what have been called "Greenscoe volcanics".

Further over the limestone is cut, and stepped back a little, by a fault, while, behind you to the south, there is a marked trough backed by another ridge of—yes, our Coniston Limestone! In fact there are several of these ridge-like exposures of limestone, for the bed has been repeated by what are known as "strike faults". The first trough is composed largely of Skiddaw Slate with occluded patches of volcanics, but exposures are not too easy to find.

Let us now return to the valley floor and the brickworks. If after leaving the main road you pass the brickworks on your right, you can follow the road to a series of quarries, the source of the brick making material for the works, and a cursory examination should suggest Skiddaw Slates. In fact quite good fossils have been found from them and the graptolites indicate an upper zone in the slates with Didymograptus bifidus. I fear the exposures are constantly changing as more and more rock is excavated, and all I can do is to tell you of some of the things I have myself seen and found. First the fossils—yes I have found them, but last time I was here the band from which I collected had been quarried away! All I can suggest is that you spend a little time looking for them in the black shale, preferably away from the lava intrusions, and you may be lucky. Now examine carefully the back walls of the various quarries for in one of them the Coniston Limestone can be seen to be lying upon "mottled shale", and as we have seen this rock usually lies at the base of the volcanics. In another, while the limestone cannot be seen actually to rest on the Skiddaw Slate (the "blue shale"), the obscured interval cannot be more than a few feet.

So here at Greenscoe we see our limestone having transgressed across the whole of the Volcanics lying closely above—if not actually down on—Skiddaw Slate, and thus I think, completing our demonstration of this great unconformity. It is true that, as usual, alternative explanations have been suggested, mostly involving faults or thrusts, and as we have seen on the crags above such faults have repeated the outcrop—but this does not necessarily affect the succession. In the last quarry near Park Farm great care is needed for here one of the limestones you can see is *not* Coniston but Carboniferous, complete with its basement bed! To find two limestones in close proximity as they are here, one Ordovician and the other at least one hundred million years younger, is to say the least both intriguing and somewhat confusing, and any explanation must involve both enormous erosion and great earth movements.

On the floor of one of the quarries last time I was here I noticed an exposure of lava with, in the cracks, some most beautiful yellow fluorspar crystals. This I suggest seems to imply the solfatara stage towards the end of this great eruption. Here I

The "conglomerate" near Sikes (Chapter 2—C).

B.V.S./Skiddaw Slate junction at the foot of Low Nest Brow (Chapter 2—C).

Dip fault—Slickensides at Appletreeworth (Chapter 3—C).

Coniston Limestone at Appletreeworth (Chapter 3—C).

Thrust plane in Bram Crag Quarry (Chapter 4—C).

Slickensided surface of the fault at the Thirlmere side of Dunmail (325127) (Chapter 4—B)

must leave the subject, and trust that the journey to this last location was thought to have been well worth-while.

References

Goodchild, J. G., 1892. "Notes on the Coniston Limestone Series", Geol. Mag., Vol. 9, pp 295-299.

Green, J. F. N., 1912. "The older palaeozoic Succession in the Duddon Estuary". Privately Published, Hayman, Christy & Lilly Ltd.

Grieve, W., 1968. "The unconformity at the base of the Coniston Group and the upward succession of the Stockdale Shales", Proc. N.E. Lancs. Geol. Assoc.

Harkness, R., 1868. "On the Coniston Group", Q. Jour. Geol. Soc., Vol. 24, pp 296-303.

Harkness, R., 1877. "On the strata and their fossils between the B.V.S. of the north of England and the Coniston Flags", Q. Jour. Geol. Soc., Vol. 33, pp 461-484.

Hartley, J. J., 1925. "The succession and structure of the Borrowdale Volcanics as developed in the area lying between the lakes of Grasmere, Windermere and Coniston", Proc. Geol. Assoc., Vol. 36, pp 202-226.

Marr, J. E., 1892. 'The Coniston Limestone Series", Geol. Mag., Vol. 9, pp. 77-110.

Mitchell, G. H., 1956. "The Borrowdale Volcanic Series of the Dunnerdale Fells", Liv. & Man. Geol. Soc., Vol. 1, pp 428-449.

Mitchell, G. H. "The Geological History of the Lake District", York. Geol. Soc., Vol. 30, pt. 4, pp 407-463.

Smith, B., 1924. "The unconformable base of the Coniston Limestone", Geol. Mag., Vol. 61, pp 163-167.

IN and around our Lake District there are many rocks which have been intruded into other rocks, and these vary in texture almost as widely as they vary in composition, from the very basic like the so called Picrites of Little Knot and Great Cockup, through mixed rocks such as those of Blae Crag in Langstrath, to more acid intrusions like the Threlkeld Micro-granite. The form of these intrusions is equally variable for, although many of the "laccolites" of former days have disappeared under further research to become recognised as stocks, dykes are many, and sills are not altogether rare. So, on at least two counts, a look at some of these intrusions may well be of interest.

Firstly then the form of the intrusion; to those to whom a dyke is no more than a drawing or illustration in a textbook, to see such intrusions in the field is very worth-while. The same can be said of a sill, and to be able to demonstrate the validity of a laccolith is something that is well worth all the time it may take in the field. In the following pages, therefore, I will try and guide your footsteps to places where many of these differing forms may be seen, while at the same time drawing your attention to not a few of our many different rock types.

Excursions:

A.   Friar's Crag

FOR a dyke we have already seen an almost text-book like one at the foot of the Dodd, but there is another readily available example at Friar's Crag and perhaps this has the advantage of not being quite so easily compared to the text-book. While proceeding towards the end of Friar's Crag, watch for the path that leads off to the left in the direction of Calf Close Bay. However, instead of going so far, continue straight on but keep to the Bay side of the crag. Here you will notice a seat for the convenience of visitors. Careful examination of the rocks here-abouts will show that two different exposures can be seen. Just above the seat a hardened form of Skiddaw Slate is exposed while above this the dolerite of the dyke is visible. At this point, unfortunately, no contact between the two rocks can be found. A little below the level of the seat, dolerite with some quartz stringers shows but following the faint path towards the end of the crag soon puts us back on to the Slates. Near the end of the crag there is a small break, or indentation in this, the eastern wall, which is caused by faults crossing the crag almost from east to west. Beyond this break, and at the end of the crag, dolerite is exposed and is heavily veined, some of these veins being quartz and some calcite. Here, on the Bay side, there is a clear cut channel between the crag and an outlying piece of rock. But this rock you will find to be different; it is Skiddaw Slate and the line of junction seems to run along this channel.

Following the dolerite brow of the crag round to its western side we soon find a much larger embayment than the one on the eastern side, but this too is due to the weakening of the rocks by faulting. The rocks here are somewhat reddened, as is often the case in the vicinity of our faults, and finding the slates isn't very difficult. But what we want is an actual junction and, pro-viding the water is not too high, this can be found by watching the rock wall low down while moving towards the boat landings. When the slates are found they seem, apart from a little bleach-ing and slight hardening, very much Skiddaw Slates and are easily distinguished from the dolerite.

So here we have a wall-like mass of dolerite that was intruded into the already existing Skiddaw Slates, and if we cannot actually see contacts all round we can in many places come with-in inches of such contacts. "Not much like my text-book!" did I hear someone say? No, definitely not! and as often as not this is the case, for while it is sometimes possible to remove the superincumbent debris to give a perfect exposure in the field, it is infinitely more simple on a piece of paper! As we have seen your text-book examples *do* exist, when recognition should present no difficulty. What is more difficult, and requires experience, is the ability to "see solid" as it is called—to be able to visualise your exposure as if most of the unwanted over-burden was removed. This ability cannot be attained in an arm-chair!

If, when you are back with your books, you care to check the word dyke (or dike, for both spellings are used) you may be surprised to find two different interpretations—first a channel, or watercourse, and secondly a wall or embankment. Curiously enough dykes of the geological kind can, and do, take both forms. Often the dyke rock proves to be much harder than the host rock and long weathering causes it to stand out, as it did at Friar's Crag. Sometimes, however, the heat of intrusion so bakes the country rock that it becomes more resistant than the intrusion, and then you get not a wall but a channel. In Lakeland parlance this often becomes a "door", like Mickledoor between the Pike and Scafell itself, the dyke here responsible (at least in part) proving to be more easily weathered than the surrounding rocks.

One of our most famous Lakeland dykes is the Armboth-Helvellyn Dyke, and as this crops out on both sides of the lovely Thirlmere valley a trip round this lake and its surroundings may well be worth-while.

Excursions:

B.   Around Thirlmere

LEAVING Keswick by the main road south, pass the exposure we have already visited at the foot of Low Nest Brow and proceed to where the new stretch of road has been made. On your right, to some extent hidden by the trees is Shoulthwaite Moss, the site of a small lakelet which has been filled in since the end of glacial times and is now occupied very largely by peat. Near the beginning of the double track road on the way south there is a quarry on the fell side, Yew Crag Quarry. The rock is massive andesite worked for use in walling and on the roads. It is worth examining for the many small faults and slip planes, some showing brecciation and others slickensiding. Along these slips there has been some mineralisation, quartz being most noticeable, but perhaps the exposure is of greatest interest because I think we have here what Professor Marr had in mind when he talked about "shatter belts"—not so much one great break in the rocks as a series of small parallel faults having much the same effect (i.e. to weaken the rock structure and to make it the more susceptible to weathering).

Proceed now past the cross roads at Legburthwaite to the road junction at the bottom of the Vale of St. John. Here from the fell side of the old road a little lane leads off to Stannah. Pass the farm buildings and the stile, to where a gate gives access to a bridge across the concrete drainage channel. On the right there is a fine exposure of andesitic lava in the form of a roche moutonnée and here, beneath a small overhang some ten feet to the right of the path, the dark stubby crystals of the pyroxine, augite, have weathered out marvellously. This is a location

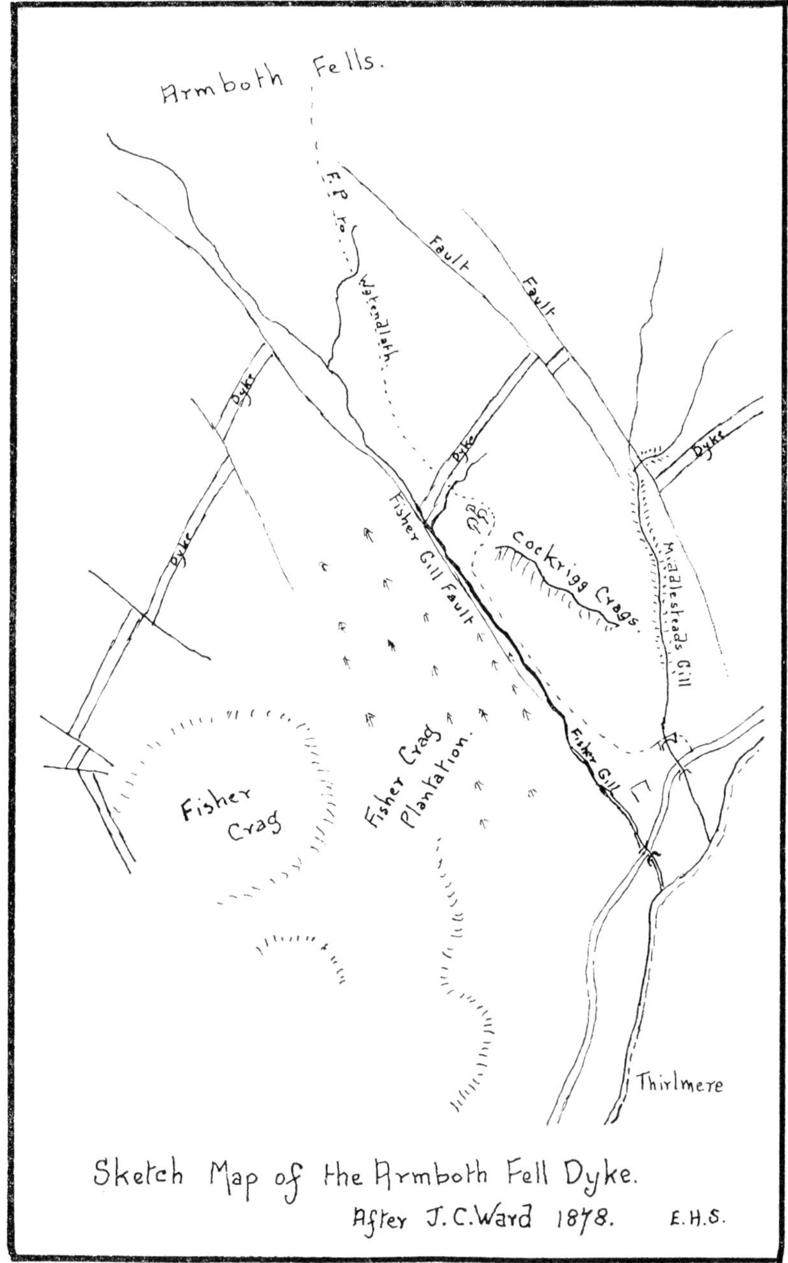

Sketch Map of the Armboth Fell Dyke.
After J. C. Ward 1878.   E.H.S.

referred to by J. F. N. Green and I draw attention to it because here we have a location and an exposure where we can talk about the mineral content of a rock without any fear of ambiguity. Augite is perhaps the commonest member of that family of rock building minerals known as pyroxine and is found most commonly in rocks of a basic character like gabbros, basalts and dolerites. Found here dominating an andesite it bestows its name to lava of this type—augite andesite. All these minerals are complex silicates of elements like calcium, magnesium, iron and aluminium, and here at any rate you can without much effort get specimens of a rather pretty rock and its dominating mineral with little possibility of mistake.

Back on the main road proceed past the King's Head and up the rise to the new car park overlooking Thirlmere. The old road which used to cross the valley between the two tarns before the reservoir was made can be seen running up towards the present road along the wall at the edge of the woods around Dale Hall. Cross the road now to the start of the Swirl Trail made by the Forestry Commission but, after passing the gate, instead of following the trail cross the stream and make for the rough path that leads up by Brown Cove Crags to Helvellyn. After crossing one or two minor streams you will come to a much bigger stream, Helvellyn Gill. This has cut a marked streamcourse with a waterfall so it can hardly be mistaken. Notice in crossing this stream that the rocks are showing a certain amount of brecciation, and then leave the path now heading for Helvellyn and follow the edge of the forest until you come to an old sheepfold with a wall running up the fellside about 30 yards beyond. Leaving this for the time being, continue until you come to a second sheepfold. This is somewhat bigger than the last and is an important landmark for it stands across a much sought-for location, the Helvellyn extension of the Armboth Dyke from across the valley. On the fell side of this sheepfold there are several large boulders lying about which are worth a glance as they show flow brecciation, and opportunities to see examples of this phenomena should not be missed as there are many forms of it. However, some 25 yards away an isolated tree grows upon the rocks—and the rock is an outcrop of the dyke.

If, like myself, you know the Armboth Fell dyke in its home location you may well find yourself a little puzzled, if not deeply disappointed, for at first sight the rock exposed here does not seem much like its name-sake. For one thing the exposure is pretty well buried in moss and lichens, and this tends to hide the rock. If, however, you pull away some of this superincumbent material what is then exposed does come at least a little nearer expectations. Tried with a hammer the results are not very re-assuring either, for instead of the pinky-brown base of our Armboth rock this seems to be of a rather nondescript glassy grey. If, swallowing our disappointment, we examine the rock a little more closely we can spot rather large pale pink felspars. They look somewhat decayed as a result of the weathering, but at least they are there and so too are the little blobs of quartz which are such a marked feature of fresh rock.

Without a sledge hammer, getting specimens at all is a matter of some difficulty and in what I could get I could find no garnets, but perhaps this is not so surprising as even on Armboth they are not all that common. Before leaving here let your eye rove up the fell towards Helvellyn and notice that on the crags above there is a marked "door". This almost rectangular corridor in the rock face is due to an extension of the dyke we have just examined, and either the dyke rock has proved softer than the Borrowdales into which it is intruded, or maybe it was better jointed and thus more easily eroded.

Still keeping parallel with the forest, follow the track which in about 150 yards crosses three small streams to come to a rocky ledge with a silver birch growing from the end. The path we have been following goes on over this bench but, leaving it to one side, go on for another fifteen paces or so to where a little juniper grows from a crevice in the rock face (approx. 322163). Casting about here you will find a wall of rock about four feet high, the face of which shows some really fine flow breccia. The most striking feature of the breccia is the size of the blocks themselves, many of which must be at least a foot across. This suggests an interesting speculation. Possibly many of you, like myself, have seen on the T.V. screen of late some of the really magnificent films about volcanoes where moving lava was shown with great chunks of consolidated lava falling down the front of the advancing stream—is what we see here the result of similar action all those hundreds of millions of years ago? While pondering this it might be worth-while to climb to the top of the bench if only for a glimpse of the magnificent panorama spread out below. Away to the left are the steep rocks of Long Crag, and down below is a peep of Thirlmere, across which you are looking down into the Wythburn valley and the road to Greenup Edge.

Return now to the sheepfold and its adjacent dyke, but instead of following the path back to the other sheepfold bear up and across the fell until you approach the wall we passed on our way up. At about 150 yards up this wall from the path below there is a very large block of rock built into the wall, and hereabouts are some most interesting exposures first mentioned by J. J. Hartley. To the north of the little crag with the prominent tree a small exposure shows really fine bedded tuff, while at a little lower level there is a pocket of tuff which seems to have infilled a hole in the rough surface of the lava. I admit that you may have to spend a little time here in order to find and check these exposures, but I think it may well be worth the trouble for their bearing on the old but still continuing controversy about the conditions under which such beds were laid down. Was it in water or, as some maintain, on dry land?

To continue our trip around Thirlmere we must now make our way back to the road and then down the lake to where, just past the little church at Wythburn, a road turns right for Armboth. Here you are looking south and up at Steel Fell, and running down to the road is one of the best median moraines in the district. Formed between the ice of Dunmail on the one hand, and the ice from the Wythburn valley on the other, it can be followed with the eye right down the fell breast. It is worthy of notice for median moraines are not so common. Right opposite this side road there is a gate that gives access to the forest. Passing through this, make your way towards the river through the trees. Here, where the stream from the Raise makes a wide bend (325127), the rock face slopes back into the bank at a steep angle. Careful examination will show marked slickensiding of the surface (i.e. a polishing and striating of the exposure due to the slipping of the rocks on the fault plane). This exposure, although to some extent moss covered, shows some fifteen to twenty feet of slickensides and a little search will show fault breccia, some of which is stained a reddish purple colour. This exposure is, so far as I know, the only one you can see of this important fault which begins between Lonscale Fell and Blencathra and runs down the Vale of St. John, the length of Thirlmere, and on to Grasmere. It is possible that the mineralisation around Tongue Gill, where haematite was once mined, is in association with this fault.

Returning now to the transport, take the road to Armboth which runs first in a westerly direction and then turns north. Notice on the hillside the ravages of ice where large exposures of tough lava have been smoothed and moulded by its passage. A little way along is Stackhow Bridge (321129), just beyond which the road passes through a rock cutting. On the eastern wall of this cutting the Wrengill Andesites are exposed and the lava is well flow-brecciated. A close examination will show that many of these fragments are coated with films of volcanic glass.

Continue to the car park at the foot of Launchy Gill. There is now a Nature Trail up the gill laid out by the Forestry Commission, and this we will follow, but first examine the rocks by the roadside. These exposures are again in Andesites but here with many pink felspars. A little further along the road, where the return path from the Trail comes down, the same lava shows a slight brecciation. The structure here, as mapped by Green, is anything but simple. He suggested a deep infold, with the Wrengill Andesite enclosing the Tuffs, and Hartley at a much later date seemed to concur, so let us follow the Trail and see what we can find for ourselves. Above Point 2 on the Trail a purple coloured rock is exposed which looks like an iron-stained tuff. Next exposed in a little gill is more tuff, but this time unstained and looking not unlike the Harrath Tuff near Millom. Above Point 4 it seems that we have passed back on to the purple tuff, while at Point 5 there are large lava blocks with garnets. Turn now towards the lake and there poised almost on a knife edge is the so-called "Clattering Stone", a most spectacular perched block clearly exemplifying its origin. No boulder this size could tumble down the mountain side and come to rest so precariously perched; on the other hand encased in melting ice it could slowly but surely settle down to be so poised.

Back at the start of the Trail, follow this lovely road along the lake in a northerly direction for about three miles to the foot of Fisher Gill where there is a rather rough lay-by (306170). Walking a few yards up the road will bring you to a signpost which says "To Watendlath". Passing the gate, you go over the old stone bridge, through a sheep pen and out into the open with, immediately before you, many erratics lying by the path. Notice that the little cairns which mark the track are made up mostly of bits of the Borrowdales, with an occasional piece of a very different rock. A pinky brown in colour, this is studded with reddish crystals of felspar and is the rock from the Armboth Fell Dyke.

As the path ascends the fellside it gets very rough—and where it begins to follow the deer fence it is more like a beck bottom—but if you watch the cairns carefully you will notice that, while pieces from the Borrowdales still pre-dominate, the bits of dyke-rock are becoming a little more frequent. Now the path begins to zig-zag around some trees to come out above on the open fell. Not far beyond these trees a small stream crosses the path close to where a prominent cairn consists almost entirely of bits of the Armboth Fell Dyke. This is very suggestive, and with a little casting around here you should find the dyke running up the bank to the north, but not making any sort of a feature, although it is at least twenty to thirty feet wide. This rather pretty rock has been described as a granophyric quartz porphyry, and examination will show well defined red felspars in a pinky-brown felsitic base. The rock contains quartz and this is rather remarkable in that it consists of bi-pyramids, rather than the more usual interstitial variety. There are occasional garnets and the rock is referred to as micro-spherulitic, though I doubt if this structure can be made out with a lens.

The deep gill to the south within the wood is Fisher Gill, and it has been eroded along a line of fault which displaces the dyke. To find the further extensions of this interesting intrusion you will have to continue up the fell and over the wall, and then some way beyond the gill. A little casting about should reveal the dyke, but at this exposure the rock has a slightly different look about it. Between here and Fisher Crag the dyke is displaced towards the lake, but on the edge of the crag it is well exposed. It is thought that these faults are tear faults—or as they are sometimes described, wrench faults—and the movement has been lateral rather than in an up or down direction. All workers have regarded this lovely dyke as an offshoot from the Threlkeld Micro-granite, although it cannot actually be traced to this granite. Certainly the chemical analysis of the two rocks as given by J. C. Ward for the dyke, and for the micro-granite by Hadfield and Whiteside, are remarkably similar:—

	Armboth Dyke	**Threlkeld Micro-granite**
Silica	67·180	67·18
Alumina	16·650	16·65
Lime	2·352	2·35
Magnesia	1·549	1·55
Potash	2·914	2·91
Soda	4·032	4·03
Ferrous Oxide	2·151	2·15
Ferric Oxide	·559	·56
Phosphoric acid	·179	
Phosphorous		
Carbonic acid	·885	
Carbon Dioxide		No deter.
Carbonaceous matter	·797	·89
Water	·753	·75

If, however, you have time for a diversion I will show you an exposure I have never seen mentioned by anyone, but one which does seem to be relevant to this suggestion. Back on the road, cross the dam to reach the main road at Legburthwaite, but instead of turning towards Keswick continue by the Forestry Nurseries towards the Vale of St. John. With Castle Rock on your right and the end of High Rigg on your left, one can hardly fail to notice the beautiful "U"-shaped valley left by the ice, nor yet, on a second glance, miss the fact that the once vertical wall is already beginning to be masked by the results of some ten thousand years of post-glacial weathering.

Continuing up the dale, keep your eye on the slopes of High Rigg and see if you can spot the displacement of the lavas by the faults that cross the Rigg—there are at least two, Sosgill and the William's beck faults. Follow the valley until you come to the turning on the left that leads to St. John's Church and to Shundraw. It is the Shundraw road we want for here, just a little way beyond the farm on the left, there is a small quarry (306236) in the Micro-granite. If you ask at the farm I don't think they will mind you going into the quarry. It is not very big and I doubt if much rock was ever got from it, but if you examine with care the quarry floor a little beyond the small building you will find it is crossed by a dyke, and the rock is not much different to look at than the Armboth Fell Dyke. It is a little more pink in colour and certainly of a finer texture, but the felspars are there, as also are the little blobs of quartz which are such a feature of the Armboth Dyke. If indeed further investigation should show that this is the self same rock then this exposure cannot be ignored, for it is here seen to cut *through* the Micro-granite, which on grounds of chemical similarity at least has often been claimed as its source.

The rock we have here been examining reminds me of something which I think most geologists acquire as time goes on—an almost unconscious awareness of rocks. Coming down the lane by Shundraw one day the bright light caught the corner of the old barn up the road, and at once I was all attention. What was that? A pinky coloured rock—well, those are common enough —but that's just it! They are *not* common in this immediate part of the country. The Armboth Dyke? Ah—yes, probably. But then instinct took over. "Probably" just wasn't good enough. Detailed examination showed that, while quite like the Dyke, it wasn't quite right . . . so where had it come from? I did not find its source that day, nor indeed for many a day, but eventually I did find it in the exposure we have just seen! Whether this sort of awareness was ever deliberately cultivated I really cannot say for now it seems to have always been with me. Perhaps, when I was very young, the stones in the walls, and even the road metal beneath my feet, were so often pointed out to me that they became second nature. I really do not know, but it has often proved to be very useful.

There is an aspect of the Armboth Dyke that we cannot pass without comment, and that is its usefulness as a "marker". The distribution of this rock is very widespread and also very interesting, and thus once seen you can hardly fail to recognise it. As Professor Marr pointed out, when the Ice Age began we had here in Lakeland a state of subdued topography which was the result of long ages of erosion. Rounded outlines and not sharp crags were the general rule. In some parts of our Lake country remnants of this old topography have survived and the area we have just been visiting is most certainly one. Think back to both sides of the Thirlmere valley, and rounded outlines are on the whole more prevalent than crags, especially so on the Helvellyn side. This takes us back to the distribution and movement of the ice during the Ice Age.

At the head of the Vale of St. John, which is a direct continuation of the Thirlmere valley, stands the Skiddaw-Blencathra group of mountains and their Carrock Fell hinterland. These were a source of ice rather than a possible outlet, but behind these mountains at the height of the Ice Age stood the ice that had descended from the Scottish Highlands. As this glacial mass at one time swept round our Lakeland mountains even into the valley of the Eden, it is not difficult to see that movement of ice from the Helvellyn range would be very restricted and would find no easy outlet due north. That movement was never very active is suggested by the still maintained subdued topography.

So what did happen to the ice that would undoubtedly form along the Helvellyn range? This is by no means an easy question and I suspect that at different times there were different answers. Perhaps one of the best studies is still the one made so long ago by Clifton Ward, even though he thought that *floating ice* had been the medium of dispersal. As Ward points out, erratics from the dyke can be traced right across the Armboth Fells, over the summit of Blaeberry Fell and down into the Vale of Keswick (and now that you know the rock try your powers of observation on the walls in the town!). In fact the rock is by no means uncommon down the West Coast. But Ward also drew attention to erratics of the dyke rock that he found round the eastern flanks of the Blencathra range up at Mosedale!

Excursions:

C.   The Threlkeld Micro-Granite

IN the early days of our Lakeland geology almost all the large granitic masses were held to be laccolites or, to use the more correct term, laccoliths. This type of intrusion was first observed in the Henry Mountains of North America and was thus described by Gilbert in 1877: "The lava of the Henry Mountains, instead of rising through all the beds of the earth's crust, stopped at a lower horizon, insinuating itself between two strata, and opening for itself a chamber by lifting all the superior beds. In this chamber it congealed, forming a massive body of trap. For this body the name laccolith (Gr. lakkos—a cistern, and lithos— a stone) will be used."

This, at the time, was indeed something new in the world of geology and, as has often been the case, it started an intense search for such structures elsewhere, including our own Lakeland. But these sudden enthusiasms have, time and again, proved little better than a trap for the unwary, and if at one time most of our granitic masses were supposed to have this form, long years of further research has shown that most of them are in fact "stocks" or on the grander scale "batholiths". The one notable exception to all this is the mass of micro-granite near Threlkeld for this intrusion does seem to have the classical form of the laccolith, so let us go and see how far we can actually demonstrate this in the field.

Let us be quite clear. These quarries, which now belong to British Roadstone, are private property and can during working hours be rather dangerous for people unaccustomed to such sites. The management is very obliging indeed and will allow visitors whenever possible, providing—and I must stress this— proper arrangements are made and a time agreed when there is no danger. Assuming that you have been granted permission and you find yourself in the main quarry, notice first the general appearance of the rock exposed. To the geologist's eye it has the look typical of an igneous rock with its lack of any true bedding and its strong joints. But perhaps this is something that one has to see time and time again before it is possible to recognise it at first sight. The rock in hand specimen is worth examination, and while it varies a little it is typically a light greenish-grey, rather fine grained rock. It consists of a felspathic base with crystals of felspar and quartz. Not much mica is to be seen and some of the felspar laths have been altered to a light green mineral. There are occasional garnets, but as usual their origin has been the subject of much debate, some authorities regarding them as original while others favour a secondary source, suggesting that they have been derived from assimilated Borrowdale rocks. If you walk around the quarry it is not unusual to see masses of both Skiddaw Slate and of Borrowdale lava that have been incorporated and in part assimilated by the intrusion. Usually referred to as the Threlkeld Micro-granite it is technically a quartz felsite, and under the microscope the quartz can be seen to be crystallised rather than interstitial—just one of the features, no doubt, that has suggested a connection between this rock and the Armboth Dyke.

Over on the eastern side of the quarry the micro-granite can be seen in contact with the Skiddaw Slates. In places the two rocks are interdigitated but show remarkably little alteration at contact. As these slates are of the dark, relatively uncleaved kind, it is well worth-while to look for fossils as graptolites have been found within inches of the contact. Many years ago now the late Gertrude Elles reported four different kinds of graptolites here and commented on their fine state of preservation. The mineral tourmaline, as the variety schorl, is not all that uncommon in the rock and might be looked for. It occurs as dark green to black radiating crystals. As faults cut the quarry face other minerals have been reported and might be sought before leaving this interesting site.

Back on the road, after thanking the quarry people for their courtesy, turn down towards the vale and continue for about a mile to where a signpost indicates the "Old road to Matterdale" (unfit for motors). Leaving the new house on the right, go through the gate and up the hill passing beneath the old tips until you come to two gates, one of them across the track. This is an iron gate while the nearby wooden one is across what was once a railway line connecting the old Hill Top Quarries with the works. It is along this old rail track on Birket Bank that the exposures we seek are to be found. Following the rather rough boulder strewn way we come to where the old track entered a cutting (approx. 321239) and the bank on the western side was walled for support. Just beyond this look for an exposure low down which for about ten feet or so has the micro-granite in contact with the Skiddaw Slates beneath: this shows the floor of the laccolith. On the other side of the cutting the slates, in places a little distorted, are clearly seen but the igneous rock above cannot with certainty be said to be in contact. Nevertheless these slates obviously cannot be far beneath the actual floor and a little digging here might well expose it. Presumably somewhere there is a pipe, or pipes, up which the intrusive matter has risen but so far none has been uncovered. In many a Lakeland intrusion advantage seems to be taken of the line of weakness presented by the junction between the Borrowdales and the underlying Skiddaw slates, and this may well have been the case here.

Return now by the way you came but, instead of going down to the road, follow the track along the fellside used to carry rock quarried at Bram Crag back to the works. You will soon come to the entrance to one of the old Hill Top quarries but as exposures are poor in this one carry on until you come to the next quarry entrance (320230). The interest here is not so much the exposures of the micro-granite as for the wide variety of rocks lying about. The first is a great boulder of flow-breccia whose surface has been well etched by the weather, while close by is a boulder of Skiddaw Slate. This has been very much altered by heat of contact; toughened and hardened as it is, it still shows quite clearly the lines of the original bedding. Since, as we saw in the main quarry, the slates at and near the contact are very little altered it seems probable that this block was a xenolith (i.e. a block that owes its hardness to its having been totally enclosed in the liquid magma). The floor of this quarry has been used as a tip for waste from Bram Crag and much of the material lying about is fault breccia from a thrust plane we will now go and see.

Follow the road south until it divides, one road going forward into Bram Crag quarry and the other dipping down the hill until, very shortly, it begins to rise again and leads to the entrance to a now disused part of Bram Crag (319219). As we found in the entrance to Hill Top, so here there are many stray blocks lying about and one which you ought to recognise is the augite andesite from Stannah. If you failed to get a specimen at Stannah that should be easily rectified here for many of the pieces lying about are very fresh and show the augites clearly. One of the commonest members of the pyroxine group of minerals, augite is fundamentally a calcium, magnesium silicate.

Lying about too are many blocks of another rock you should have little difficulty in recognising, the Purple Breccia so common in the lower reaches of Cat Gill and along the eastern shores of the lake. One interesting feature of the bits here, however, is that many of them carry occluded bits of dark Skiddaw Slate. Thus you can carry away a hand specimen which clearly shows that the vulcanicity of the Lake District began by blasting a way through Skiddaw Slates! Not uncommon either are blocks across which run peculiar parallel lines, and these should remind you of the fault face we saw at the foot of Dunmail with its slickensiding. This polishing of the rock face as a result of fault movement is very common in this quarry, as we shall see shortly.

Meanwhile, moving into the entrance cutting notice that on either side, but perhaps best displayed on the eastern wall, there seem to be two distinct types of rock visible. Nearest to the entrance, bedded rocks which appear to be up at a high angle can be seen. These are Skiddaw Slates pushed up out of line by the intruded micro-granite with which they are in contact. If the slates are disturbed they again show little signs of alteration by heat of contact. On the eastern wall of the quarry not many yards from this contact there is a large patch of slickensiding, while above this, and running the length of this wall of the quarry, there is a distinct break in the rock. This is a low angled fault, or thrust plane, such as would undoubtedly have delighted Prof. Marr! The rock, both above and below the thrust, is the micro-granite, but along the line of the break much alteration can be seen with white quartz in evidence. At the northern end of the quarry the brecciated rock along this thrust is very thick, and this is the reason why this part of the quarry has been abandoned. When a rock is worked very largely for roadstone, as it is here, then its crushing strength is a very important item. But if masses of this fault breccia were to be included then crushing strength would become altogether too variable, and would lead to trouble with the purchaser. To quarry the rock and attempt to separate out the fault breccia would be much too costly. So notice how this trouble has been overcome. At the northern end of the quarry there is a marked step with a continuation beyond, but with the offending thrust below this part of the quarry's floor! Problems like this are often the province of the practising geologist.

Fault breccia, much like the one we saw up on Skiddaw Dodd, lies about the quarry floor, but here the angular fragments are pieces of the micro-granite. This breccia also shows "vugs" but the enclosed quartz crystals here are often beautifully clear and well formed, showing both prism and pyramid. Besides these quartz crystals little cubic crystals of both iron and copper pyrites occur. Since calcite too can be found along with odd traces of galena it would seem that the mineral infiltration took place over a long time and possibly at more than one temperature. Many of the blocks lying about are well slickensided and small specimens clearly showing this evidence of fault movement are not too difficult to obtain. Thus you can carry away with you from here evidence of many geological happenings that took place hundreds of millions of years ago.

If the line of weakness of this thrust once gave passage to mineralising solutions it now gives rise to many trickles of water, thus providing a useful illustration of the fact that many rocks, while in themselves impervious, can still be water bearing, providing that they are strongly jointed as the micro-granite is here. This infiltrating water is in large measure the cause of the predominantly brown colour of the rock along the line of break, as distinct from the white quartz. If the mineralising solutions deposited the pyrites, these sulphides have been acted upon by the percolating water to produce hydrated oxides of iron to discolour the rock.

I have often been asked—what is the throw of the fault, and by how much has one limb of the thrust over-ridden the other? Frankly, from what can be seen this is an impossible question to answer, for both sides of the slip plane are essentially the same rock and thus there is nothing to indicate the amount of slip. Another very interesting but also rather awkward one is—what is the age of the movement? Obviously this post-dates the Ordovician vulcanicity but by how much is at present a moot point. (The age of the Micro-granite itself has long been debated. Current work suggests an age of 445 $\pm$ 15 M.Ys., roughly concurrent with the Borrowdales.)

When you have had your fill of this most interesting quarry make your way down to the public road and walk back towards your starting point, the old Matterdale road. As you walk along you might notice the marked evidence of glaciation. First there is the flat valley floor suggesting that in post-glacial times we had here a shallow sheet of water which was more or less rapidly filled in by loose detritus washed down from the hillsides. The hog-backed little mound with the small building is a drumlin aligned with the valley, while at right-angles to the run of valley there is another long mound through which the road has been cut at Lowthwaite (318225). This is an esker, the sandy infilling of a sub-glacial channel. Do not fail to notice straight ahead the really fine truncated spurs of Blencathra which clearly indicate that at some time the run of the ice was *across* the head of the Vale of St. John. The marked bands of B.V.S. stand out on High Rigg where the grassy slopes between the little cliffs due to the lava are caused by bands of ash, or sometimes the more easily weathered flow-brecciated tops of the lava flows. The eastern side of the valley fails to show any comparable structure, and this failure of continuity is one of the indications of the great fault that runs down the length of the valley.

A little way beyond the old Matterdale road turn off to the west and cross the bridge over what used to be, long ago, the River Bure. Continue to where the road branches again. Leave the road that runs up to Shundraw on your right and, following the one to the Church of St. John, you will come to the last cottage by the roadside with, just beyond, a gate and an old barn. A few yards beyond this gate there are outcrops by the side of the road and a tap will show exposures of micro-granite (308229). At the next bend outcrops can be seen beyond the trees in the field which investigation will show to be the same rock. In several places here, however, the rock has a peculiar banded structure, very suggestive of bedding, while further over in the bed of the stream, despite the heavy cover of glacial drift, undoubted Skiddaw Slates are exposed and these cannot be far below the floor of the intrusion (see map page 49).

Continue now to the new Youth Centre run by the Church and to the stile over the wall immediately opposite. Taking this path, slope a little to the west as you continue up the rising ground to where a slight depression breaks the rise. If you tap the rocks hereabouts you will again find the micro-granite. Continuing still in a westerly direction towards the wall you should find small exposures of what is undoubtedly Skiddaw Slate (304225), while as you go further up the hill you will find many exposures that are not so easy to interpret. Careful examination will show distinct signs of bedding, even more markedly so than the exposure we saw in the field on our way up. It is quite true that as a hand specimen this rock would be extremely difficult to place for it is both hard and tough. Even on a freshly broken surface it does not look very much like the Skiddaw Slate to which we are accustomed. Alteration here has been very considerable, both by baking and the incorporation of actual magma, making the rock anything but easy to recognise. This clearly illustrates the necessity when dealing with rocks to take into account not just the rock as such but its "field relations", and this needless to say just cannot be done in a study or laboratory.

Here on Skelthwaite Crag let us sum up the evidence regarding the form of the intrusion of this micro-granite. First in the main quarry we saw a junction between the Skiddaw Slate and the igneous rock to the east of the mass. Then in the old tram track on Birkett Bank we saw an exposure of the slate, on one side overlain by and in contact with the intrusion, and on the other side obviously not far below it. In Bram Crag we saw the southern contact between upended slates and the micro-granite. Lastly, here on Low Rigg, we have seen much altered and partly incorporated slates forming what must have been the roof of this most interesting intrusion. This can be paralleled by exposures on Clough Head where the rocks above the main mass simply grade into the micro-granite in such a way that Ward felt unable to clearly distinguish any "junction" at all! I chose the roof exposure on Low Rigg simply because the exposures are more accessible, while being comparable with the ones on Clough Head. The evidence we have seen, I suggest, clearly indicates the form of the intrusion—a laccolith. The Low Rigg exposure is some 150 to 200 feet lower than the Threlkeld mass on the other side of the valley, which suggests that we are here on the downthrow side of the fault.

Hadfield and Whiteside did express some doubt as to the exact form of the intrusion here on Low Rigg, suggesting the alternative of a sill. The general form seems nearer a laccolith than a sill, and certain it is that before the road was tarmacadamed near the church Skiddaw Slates were visible along the track. The Threlkeld mass is now generally agreed to be of this form so if you want to be sure, and have the energy of youth, try Clough Head, where at the same time you can seek specimens of volcanics welded to Skiddaw Slate.

Excursions:
D.   Bassenthwaite... and beyond

TAKING the road, which by now will have become familiar, along the Skiddaw side of Bassenthwaite Lake, pass Mirehouse and the entrance to the Dodd forest road to come in about 1½ miles to Spout House. Here a side road leaves the main road on the right, and after some 500 yards passes an old quarry which has been made into a lay-by. Just above this a gate (236311) gives access to the fell, and immediately through the gate a small stream exposes the Skiddaw Slates, thin bedded and apparently up on end. Following this stream up you will pass two concrete water supply tanks to come to a large tree where the stream makes a bend. From here make for a gate which, looking up and across the field, can be seen in the wall. A rough track then leads across a field to another gate with Barkbeth Hill on your left and Watches on your right. This old track is much better defined now and following it you will come to yet another gate above Southerndale Beck. Continue along this track, but notice that immediately through the gate there are exposures of dark, thinly bedded Skiddaw Slates. You pass many such exposures along the fellside as the track dips down to the stream, a sheepfold and a rough bridge across the beck (243304).

Looking up towards the east from here is the object of our excursions, Little Knott and its Picrite. First recorded by that keen observer John Clifton Ward, this intrusive rock was described by him as a diorite and it provides one of the prettiest of his many coloured microscopic rock sections. In 1885 Bonney described the rock as a picrite, and certainly it is one of the more basic of Lakeland's many intrusions. The word "picrite" means bitter, a reference to magnesia and its compounds which have a rather bitter taste, and to the fact that magnesia makes up some ten to twelve per-cent. of the rock.

But now to find the rock that brought us here and I must confess to some puzzlement. Always it has been referred to in geological literature as the Picrite of Little Knott, but cast your eye up the hill and I think you will agree that exposures are not very obvious. Now turn about and look west. A little way up the fell side, across and to the right, there is a bold exposure above brown coloured scree that will immediately suggest "igneous" to those of you who have acquired the geologist's eye. All the way up to the summit of the fell there are exposures which certainly do not in the least look like Skiddaw Slates, but this hill is Watches! Why this confusion I do not know unless, as does often happen, exposures have altered radically in the intervening years. However, let us cross the stream and follow the track up the slopes of Little Knott.

The eye is first attracted by a pile of boulders and, assuming something a little heftier than the coal hammer is available, we will break off a piece. A grey-green rock of rather fine texture, this is the picrite, but there is no actual exposure. Next to a large flat slab is another boulder, which has weathered to a rich brown even when broken across, thus suggesting the high iron content of the rock. At the summit the only rocks exposed are Skiddaw Slates but they are tough and hard, which suggests that although no picrite can be seen it cannot be far away. At the southern end of this exposure the rock is almost indeterminate, only the remnants of bedding giving it away, but this you may remember we have seen before in connection with the Micro-granite. Some 35 paces towards the south a rock with a more slabby aspect is exposed, and this is the picrite. The rock here is badly weathered and no actual contact is to be seen, although but a few paces away well baked slates are exposed. For the rest a thick cover of drift hides the rock.

Making our way back to the sheepfold we can head for the prominent rock surrounded at its base by scree. A tap here shows the same grey-green rather fine textured rock we have seen before. Now make your way up the exposure on the side of the grass covered hollow (242303). After some 25 paces or so, well baked slates can be seen, and although the actual contact is hidden by the scree it obviously cannot be far away. Tapping the picrite as we make our way up the fell side will show the rock getting progressively coarser and much darker in colour. Probably our grey-green fine textured variety came close to a chilled contact, while here with slower cooling crystals have had chance to grow. The large hornblende crystals, in various shades from black to green, can be anything up to 5mm in length and under the lens show striations. The groundmass is variable but is often a greenish-grey where most of the original minerals have been replaced by alteration products. Pyrites grains are fairly common, along with dark specks of magnetite, while the greenish flakes are most probably chlorite after biotite. Augite, olivine and enstatite have all been suggested as original minerals, but none of these can be identified in a hand specimen here. Two technical terms are applied to this rock: it is called melanocratic —meaning a dark coloured igneous rock, while its minerals themselves are often referred to as mafic—i.e., dark coloured ferro-magnesian minerals.

We have observed that the rocks around this intrusion are

well baked by the heat of contact, and this can be seen to perfection at the top of the fell, where at the south-west corner overlooking the lake an actual junction can be examined. Besides the many exposures here on Watches, and the ones we saw on Little Knott, there are two others further to the east around Dash and Great Cockup. Are they all part of the same intrusion and inter-connected? It is thought so from the comparable nature of the rock at all of these exposures, but the heavy drift cover makes it impossible fully to demonstrate this.

Two points of interest there are in connection with the picrite —its age and its source. In and around the central mass of the Lake District there are a great many intrusions of a widely vary-ing character. Most common in the older rocks, they are not so common in rocks of Carboniferous or younger age, and while this is suggestive it is not entirely satisfactory. Here we could do with help, and in these days of radio-metric dating the first question should not be too difficult to answer. Given the age we could at least make a shrewd guess as to the source, but for the moment all I can suggest is that most cases of large acid intru-sions around our district seem to have been preceded by a basic phase. Basic rocks are widespread in association with the Butter-mere and Ennerdale granophyre, and the evidence seems to point to the gabbro of Carrock Fell having been emplaced before the granophyre. So the granite of Skiddaw itself can be considered as at least a likely source. Before leaving this interesting area a little attention to the many exposures of Skiddaw Slate on the side of Watches overlooking the lake may well be worth-while, for graptolites in a quite good state of preservation have been found. *Didymograptus nitidus* and *Glyptograptus dentatus* have both been reported.

Regaining the main road continue towards Castle Inn, but at the cross roads by Bassenthwaite Church turn right to pass through the village itself. Here a left turn at the far end brings you on to a country lane which runs along the fell side to Bassen-fell and Castle Inn. If you take this lane leisurely—as all such lanes should be taken—you can admire the fine views, but don't neglect to keep your eye upon the fellside too, for above the little group of houses and farms known as North Row you should spot an interesting old quarry. Just inside the entrance drive to a new house an old gate stands across the grass grown track that leads to this quarry (222324). The place itself, I fear, is an awful tangle of briars and undergrowth, but the back wall shows some sixty feet or so of a dark grey smooth rock, the fine texture of which is reminiscent of the dolerite of Castlehead. Along the length of the quarry, Skiddaw Slates can be seen to overly the igneous outcrop and the form of this intrusion is a sill. In the recently published Survey Memoir, however, the rock is described as nearer to the Picrite of Little Knott and part of a widely distributed suite of rocks of a basic character.

At Castle Inn turn left at the cross roads to run down to Bassenthwaite and Ouse Bridge. Crossing this lovely stretch of the Derwent as it leaves the lake, turn right in the direction of the main Keswick to Cockermouth road continuing to where, within 50 yards or so of the road junction at Brathay Hill, a side road doubles back to Crag. Follow this lane past the farm of Low Crag and the houses at Crag (190313) to where it looks as if the lane is about to peter out. In fact, very narrow though it is, it does continue to join yet another lane further up the hill. However, proceed cautiously until, about where the lane begins to rise, there is rock showing in the bank on the right beneath the trees. This is another interesting rock but the real difficulty is to get an unweathered specimen. Penetration has been rather deep and, as the iron content is high, decay tends to bring about discolouration. The rock has been described as a spessartite, and felspars can sometimes be made out along with darker coloured minerals. This is considered to be yet another member of the rock group which we saw at the last exposure and it is also thought to be related to the Embleton rock which we shall see later.

However, I have to admit the name troubles me. "Spessartite" has for well over one hundred years been the name applied to a species of manganese bearing garnet, so called after the part of Bavaria where it was first found, so why after all this time add confusion by introducing a new application? One of the most common objections to our science is its multiplicity of names and long words, and in this last few decades this has tended to get worse. Even I, who was reared on geology and its literature, find some modern papers so irritating as to be almost unreadable. You add nothing at all to "right handed" or "left handed" by talking of "dextral" and "sinistral"—if all you mean is right handed why not say so? If we are to win back again the respect and interest our science enjoyed in the early years of this century, then the time has come for a long hard look at ourselves! We might all read, or read again with profit, that last address of the late Professor Read in which he spoke so scathingly of "those who know more and more ... about less and less," and wherein he made such a strong plea not only for clearer thinking but for equally clear and jargonless expression.

Meanwhile here we are in a little narrow lane where, unless we want to become unpopular, we must not stay too long, so let us to our exposure. Towards the left the igneous rock is split by a band of Skiddaw Slate, while a little further over the same slate can be seen to form the wall of the intrusion. As to the exact form taken by this intrusion not enough could be seen last time I was here to form any definite conclusion, but it looked rather like a dyke. If you have broken away quite a bit of rock in your endeavour to get a fresh piece this is definitely one of those places where it should *not* be left lying about!

Having tidied up, proceed with care up the hill until you come to the next lane, where turn right to pass an entrance to Higham School and a staff bungalow before reaching yet another cross roads. Here turn left and note that, just before the end of the wood, there is a rather overgrown entrance to an old quarry (183315). Skirting the usually wet floor keep close in to the side by the road where you will find an exposure in the bank. Bands of Skiddaw Slate sandwich in a tough hard grey rock which takes the form of a sill. All the rocks exposed in this bank are up at an angle but you can see that the intrusion follows the bedding planes of the host rock. Described as a diorite the rock is unrepresentative of its class, and even the rock on the back wall of the quarry to which it can be traced is not much like the diorite we shall see at Embleton. However, it is not so much the rock type that is of importance here as the form of the intrusion, and while it takes a little careful checking I trust this can be traced out.

Continue along this road in the direction of Cockermouth until, approaching Rake Wood, a line of pylons for the grid cross the road. Away over the fields to the north, and near the foot of one of the pylons, there is an old quarry (173319). Here you can see arched up Skiddaw Grits with an intrusion of diorite which has failed to break through the cover.

Still following the road until you have left the forest behind and begin to approach the main Cockermouth road, notice a ridge on the right which is Watch Hill. Almost on the skyline is an old quarry (149317) with exposures of the famous and long debated grit. This itself shows, rather remarkably, columnar jointing, while other exposures you can find are undoubtedly igneous but of somewhat obscure relationship. Felspars are visible along with darker minerals, and in places vesicles. The grit, over the years, has been considered to be of many different ages, even being correlated by Green with the Caradocian, but in recent years the discovery of fossils in the underlying beds has enabled the Survey to place it with the Loweswater Flags. Not all the questions are answered, however, for some parts of this grit contain pebbles that are igneous, mostly of granophyre or rhyolite, the source of which is something of a mystery.

Where the road you have been following meets the main road turn sharp left away from Cockermouth and, passing the bridge with its abrupt right turn, continue up the hill to a large lay-by on the right hand side of the road. Immediately opposite, across the fields, there are outcrops on the side of the hill known as

Slate Fell. If you walk up the road to the top of the hill a stile gives access to a footpath across the fell. Here you can see a number of exposures of a dioritic rock emplaced in hardened Skiddaw Slates (146305). Doubtless the igneous exposures are all inter-connected underground and both rocks have, at one time or another, been quarried. The largest outcrop—the one further away from the road—is a light grey in colour, is vesicular in places and does show at least some signs of crystal structure. This mass runs parallel with the strike of the Slates and while the other exposures are more irregular it is probable that all are in the nature of sills. To the north-west the intrusion is cut off by a fault.

Back on the road notice that directly opposite the stile is a lane. Following this you will come, in about 150 paces, to a rather broken down gate on the right which gives access to a grass grown track that wends it way up the hill. The first quarry (152306) running off from this track shows nothing but hardened and somewhat whitened Skiddaw Slates; thin bedded and rather silty as they are you can see that they could have been used for roofing material. Although no igneous rock can be seen in this quarry it is pretty obvious that it cannot be far away.

Continue now up the hill to a second quarry. At the entrance more hardened and bleached slates can be seen on the left and the exposure runs back to a rock wall that has a somewhat different look. Massive, and without any obvious bedding, it proves hard and tough to the hammer. A grass grown patch is followed by a vertical face that is beautifully polished and slickensided, with the striations clearly indicating that movement has been lateral, not up or down. So here we have a fine example of a tear fault, or as this kind of fault is often described, a wrench fault (see page 117). I am pleased to pin-point this exposure for I know that, although text books often give the impression that this sort of thing can be seen quite commonly, in fact even in and around the Lake District where we have a great many faults there are not many places where you can see this type so well displayed. About half way along this wall there is something else well displayed, for here a thick slab of slate forming one cheek of the fault has been left clinging to the fault face. Lying about on the quarry floor, which is of Skiddaw Slate, there are many blocks and boulders. Some of these (there is one right in the entrance to the quarry) are of diorite, again a blue-grey rather fine grained rock but it does show some crystals. However, diorite seems to be in the minority for most of the rocks are of Borrowdale Volcanic Series, but just how they come to be here in such numbers is something I cannot explain!

Back on the main road there is another quarry just over the crown of the hill. The rock exposed is mostly badly weathered diorite but a wedge of slate does run across the quarry. On the left much debris lies about from a fault which cuts the rocks, and the late T. Eastwood recorded sphalerite from it, but as the Cockermouth Memoir records the findings of some 35 years ago perhaps it is not surprising that no trace can be seen now. It is a great pity that the interruption of the Second World War delayed for so long the publication of this very useful work.

If you continue now as if you were heading for Bassenthwaite and Keswick you will soon come to the little hamlet of Embleton. Passing the Blue Bell on the right, watch out for the village post-office with the Wheatsheaf Inn opposite. A rough track turns off the road by the side of the inn to run up the hill past a new bungalow. If you follow this track on foot you will soon come to the entrance to the Seathwaite Howe Quarry (175307), with a really fine view of the Embleton valley at your feet and over to Sale Fell across the valley to the south. To the right on this little plateau stands the remains of what was once the quarry powder house (recently destroyed by vandals), and behind it a rock wall shows very badly crumpled Skiddaw Slates with a fine, almost hair-like fault running down the middle.

But now to the quarry itself. An isolated knob of rock divides the entrance into two, and entering by the one nearest the wall we have just examined you can see Slates overlain by a different rock—one with an igneous look, in fact the diorite. Apart from bleaching not much alteration to the Slates is apparent. Passing into the quarry we see it has been worked on two levels. The older floor runs like a collar around the lower level, and if we follow the eastern wall we shall see why, many years ago now, the quarry was abandoned. Lying about there are many great blocks of diorite with a characteristic speckled appearance. The lath-like felspars are mostly decomposed, and interstitial quartz can be seen among the felspars, hornblende and mica. Some of this quartz is actually intergrown with the felspar and there is a little pyrites and some calcite. The rock texture will be found to vary considerably in different parts of the quarry. But not all the blocks lying around are of diorite for many are obviously Skiddaw Slate, and looking up the wall of the quarry you can see slate peeping through the igneous rock. Across on the north-western wall a considerable thickness of overlying slates, quite well bedded, can be seen.

It is evident that while the general form of this intrusion is a sill its outline is very irregular and the thickening of the over-lying slates to the west, the "overburden" as the quarry-man would say, made further exploitation in that direction uneconomic. To counter this they went down to a lower level, but if we go down now we shall find that, while good rock is exposed in the eastern wall, the northern wall looks different. Here it would seem that the intruded magma has picked up masses of slate giving a very indifferent rock for the quarryman's purpose. Unfortunately it was about this time that specifications for road material were becoming much more stringent, and I suppose that these troubles led to the closing of the quarry. As the quarry was owned by the same people who worked the Threlkeld Quarries, it was probably thought more economic to concentrate effort in one place. There is just one point to notice before we leave the quarry; cast about until you can find an exposure which clearly shows the dip of the beds and you will see that the slates dip in towards the hillside, i.e. in a north to north-westerly direction.

On regaining the main road take the minor road almost opposite that leads by Lambfoot to Wythop Mill. Having crossed where once a wiser generation maintained a railway, you will come to what for many hundreds of years was a sleepy out of the world little hamlet, a few old-world cottages clustered around the mill. But all this is now changing for a group of fashionable residences is springing up, made possible by the motor car! As you come out at the cross roads by the bridge you have a choice. You can go straight forward and follow the stream until stopped by a gate across the road. Close by this gate, on the left, there is a small quarry (190292) which exposes one of the many dolerite dykes around here. Leaving the car here you can go through the gate and, turning left, follow the old track up the hill to the shoulder of Dodd Crag. It is a rather steep ascent and, if you wish an easier though longer way, instead of going across the road make the awkward turn left and continue along this undulating fellside road to Wythop Church (190301). Here there is plenty of parking space on any other day but Sunday.

Going through the churchyard you will find, by keeping the church on your right, a stile in the corner under the trees. A second stile gives access to a rough grass grown track which slopes up the hillside. Passing the gate across this track you can, if you are short of time but not of energy, turn here and make for the top of Sale Fell direct. If, like myself, you prefer to take things more easily, giving yourself time to drink in all that can be seen around, you will continue up the track. Near the top of the rise, and some 30 paces or so over to the right of the track, there is a small but interesting exposure. The outcrop is in that part of the Skiddaw Slates known as the Loweswater Flags, rather sandy or silty beds with intercalations of shale. Here a strong bed of sandstone shows a distinct curve—a shallow arch or anticline. This is of interest because the many other exposures of the Slates we shall see all show strong dips into the fell, roughly S.S.E. This anticline runs through this side of Sale Fell parallel with a syncline on the other side of the valley, the southern limb of which gives a dip in the opposite direction.

The track now bends towards the left and where it meets a stone wall we can leave it and head easterly along the fell. Having left Dodd Crag on your right you will find a series of heaps of stone all of which show much quartz. The rocks themselves, where they show through the drift, have many quartz stringers and one exposure shows a wall of rock several feet high with more quartz than slate! This suggests that the intrusion we are seeking, the Sale Fell Minette, cannot be far below the surface. Looking towards the Skiddaw Range you can see a corner of Bassenthwaite Lake on your left, and up on the skyline a bold rocky outcrop. This is the minette. As the actual exposure is small, and the fell top extensive, one way to pinpoint this place is to make for the cairn on top of Sale Fell. Here if you turn west you can hardly miss an old quarry (193296) in the slates: the minette exposure is over the shoulder from here on the valley side. The little quarry itself is of interest as the beds show clearly both bedding and cleavage, with some taking the cleavage and others not.

The first thing to notice about the minette exposure is its general appearance. All the many exposures we have seen on the way up have clearly shown bedding, often with a high dip into the fell. Here all we have is a series of blocks that look as if they have been tumbled together—no signs of bedding, no dip, so on this count alone we can expect an igneous rock. If you break off a piece you will find one of Lakeland's most attractive rocks. Dark red in colour and of a medium grain, the most conspicuous constituent is a green chlorite which has almost entirely replaced the original biotite. The pink felspar is mostly plagioclase. A potash felspar is present, but as this is in perthitic intergrowth with the plagioclase, you would need a thin section and a petrological microscope to demonstrate it. When felspars intergrow the structure is described as perthitic, but when they intergrow with quartz this is described as granophyric.

All the various intrusions we have visited are regarded by the petrologists of the Survey as different variants of one intrusive magma, with the picrites of Little Knott and Dash at the basic end, and the so called "Embleton Granite", really a quartz diorite, at the other. The prevalence of biotite in most of these rocks, or as we see here, chlorite after biotite, suggests the group of intrusives known as lamprophyres, and these are most often to be found in association with granitic masses. Around the Shap Granite, dykes of this type radiate for miles in all directions and can be seen as far away as Knock Beck near Appleby, while a dyke very similar in appearance crosses the floor of a quarry (363306) behind the new bungalow due north of the church at Mungrisedale.

Besides the interest of this pretty rock itself, for minettes are by no means common, there is the light it throws upon the glaciation of these parts—or perhaps I ought to have written the question it poses. From this comparatively small exposure the ice in its passage carried pieces away, but all these erratics have been found to the *south* of Sale Fell! In Ward's time the conception of an ice sheet had not been formulated and the distribution of erratics (and Ward is still one of our greatest authorities on this!) was regarded as due, either to valley glaciers, or to submergence at the close of the ice age beneath the sea when some were distributed by floating ice. Since Ward's time, of course, the submergence theory has been completely abandoned and an over-riding ice sheet accepted. Since distribution by valley glaciers completely fails to account for Sale Fell erratics and their distribution, it was at one time thought that the Scottish ice sheet was responsible. The great objection to this theory is that while Galloway erratics are by no means uncommon to the north of Sale Fell and the Embleton Valley, none has ever been found along with our minette erratics to the south. It would seem that a likely explanation is that our Lakeland ice, trying to find its way from the central mass to the sea, met the Scottish ice sheet hereabouts and was turned southwards across the fells by it.

Considering the Ice Age we are admirably situated up here for a few words about the valley at our feet. To our right we can see a corner of Bassenthwaite Lake with the river Derwent leaving it to wind its way behind the fells of Elva Plain opposite. Down below us stretching back from the lake is a wet valley floor, with the name of the little cluster of white houses across the valley giving point to the not so long ago—Dub Wath, the ford across the wet place. Looking now to the west we see the long trench of the Embleton Valley, with at its western end a stretch of hummocky moraines. Along the floor of this valley today runs a stream which could not have cut it; it is a "misfit" stream following the path of the pre-glacial river Derwent. Probably while the valley was still choked by stagnant ice the gritty meltwaters from our Lakeland ice began to cut the river's present course behind Elva Plain and, once initiated, this new path was persisted in, for to find its old outlet the valley beneath us would need to be flooded to a considerable depth. I am not sure that the feature I have in mind is really well seen from here, but if you run your eye along the fell side from Seathwaite Howe to Slate Fell there is an obvious gap. We came through it on our way to Embleton and, at least when seen from a little further west, this little col has a perfect "ice line", clearly demonstrating the passage of ice. This is by no means an uncommon feature in our Lakeland.

If on your way down you care to head for the lake and descend by the foresty road as members of the Cumberland Geological Society did recently, you may, like some of our keen eyed youngsters, spot bits of sphalerite, zinc ore lying about the path. And thereby hangs a tale! Not long ago a team of prospectors using the micro technique took their samples all along the fell side from the streams and decided after a check on parts per million of the minerals contained that here-abouts was the ideal place to prospect for zinc! What they didn't know was that the Forestry Commission bought thousands of tons of the tippings from Force Crag Mine for making roads in the forest! As one of our Cumbrian experts said to me with a grin, "They nivver ask the locals!"

In my introductory remarks I did say that, apart from locations which would fit into the excursion pattern, I would draw attention to other places worth visiting. This subject of faults is such a case, for while we have a great many in Lakeland the type we have seen earlier today is not often so well displayed. However, it so happens that one of Professor Marr's "shatter belts" is a fault of this type (i.e. a fault where movement has been more in the horizontal than in a vertical direction), and as this is one of the major lines of faulting in the Lake District its location is worth pointing out.

Besides the great north-south line of faulting, some evidence for which we saw near Dunmail Raise, there is a major fault line which runs almost east-west across the district by way of Langdale, Rosset Gill, Esk Hause, Sty Head and the gap between Great and Green Gable to run on down Ennerdale. Along its length there are many places where shattered rock, for the most part stained red, can be seen, but in Rosset Gill (249074) beneath Rosset Pike the rocks are both stained red by haematite and well brecciated. Here too, on the north wall of the gill, a nearly vertical fault plane shows strong horizontal striations, demonstrating lateral movement along a wrench fault. J. J. Hartley in his "Volcanic and other igneous rocks of Great and Little Langdale" (Proc. Geol. Assoc., 1932) says that this same fault crosses the Blae Tarn col to run out up a tree lined gully on Lingmoor, but this extension I don't know. If you are in this part of Lakeland a check on both locations could be of great interest.

Excursions:
E.   Langstrath and Blae Crag

WHILE the ultimate object of this excursion is to examine
that remarkable igneous complex of Blae Crag, one can
hardly do this and at the same time ignore our surroundings
for we are about to venture into what is without doubt one of
Lakeland's little gems. In going down Borrowdale one cannot
fail to notice that just beyond Rosthwaite the road is hemmed
in between the mound of the Howe and the river. This great
mound is just one of a ring of moraines that cross the valley here.
At Stonethwaite the church is built on such a mound, a continuation of the ring which marks a long continued halting place
for the great glacier that once swept down Langstrath.

But now to Stonethwaite, locally "Staenthut" meaning
"the stony field". In the days of my youth one of my greatly
admired authors was W. G. Collingwood, and of Stonethwaite
he has this to say:— "Perhaps the most accessible and beautiful
scrap of real antique landscape, domestic as well as mountain-
ous, is Stonethwaite in Borrowdale and the length of rude
ancient road leading to Langstrath. There one really sees what
the poets and artists of old saw—the tumble of ragged wood from
the crags above, the rush of a rocky stream below, and the
winding, through little intakes, of a rural track, useless for
wheels but glorious for the feet of the faithful." Well, if you have
the feet of the faithful follow me, and while not forgetting to
take our hammers we will "take a glimpse of this Paradise
earthly". Beyond the little hamlet you may not go, on wheels
that is, and if your visit is in the summer time you may be well
advised to leave your wheels at the first available site and continue on foot. Almost unspoilt so far, I can only hope that
Stonethwaite may long so remain for if the hamlet is a little out
of the world today it is only what it has always been. But this
does not mean that it has no story, far from it, for once of a day
it was a bone of bitter contention.

In Norman days much of Lakeland was granted to the monasteries that sprang up in a ring around the mountain fastnesses
into which the Norman lords felt it was wiser not to intrude with
their arms. After all were not the dalesmen blood brothers,
quite capable of looking after themselves, and especially so on
a terrain they knew so well? So by peaceful penetration our
dales were to be brought within the fold. Borrowdale was

divided up between Furness Abbey on the coast and Fountains down in the Yorkshire Dales. How Fountains came to possess part of Borrowdale is a long and rather sad story, but in 1211 the two abbeys made an agreement about their mutual boundaries. Unfortunately our Stonethwaite seems to have been missed out of this agreement, and at the time seems to have been considered of little importance. But as the years passed, and under the careful husbandry of the monks our little place became a famous "vaccary", or dairy farm, and the two abbeys began to squabble as to whom it really belonged. Furness said it was and always had been hers. Fountains maintained it had been assigned to her, and it was her management that had made the place. Whether there be a moral in the tale for these days or not I do not know, but the case was submitted to one set of arbitrators after another without any sort of agreement being reached. So then the custodian of Cockermouth Castle wrote to the king, complaining that decisions about the land were being taken without consultation of any sort with him! This was clearly illegal, so the king took our little vaccary under his own wing!

But the monks of Fountains were quite equal to that little emergency. They invited the king to a deer-hunt, at the happy conclusion of which they dined him and wined him—and then offered forty shillings to have the small pasture in Borrowdale and got it! At the dissolution Stonethwaite passed back into the hands of the Crown and so remained until the reign of James I when these Crown Lands were sold and the Great Deed of Borrowdale was drawn up. One of the things of interest about the Deed was that with very few exceptions the names were the same as those returned by the monasteries before the dissolution. One of these exceptions was Daniel Heckstetter, Gentleman, of Rosthwaite, and he was the son of the original Daniel, the leader of the German miners who came here in Elizabeth's day.

I trust I may be forgiven my diversion into the past of this little place, but as we pass through we can at least think back to some of the interesting things that happened here long ago. Taking the old road that Collingwood mentions, a road that can be but little different now to what it was in the days of the monks, we will head up the dale. As we are leaving Stonethwaite notice a new white bungalow which, so says a notice, holiday people may hire. Our lane wends behind this place and on the bank to the right is our first rock exposure. In this part of the valley most of the exposures are in the volcanics and this proves to be a lava. It is rather a nondescript one perhaps, for it shows little structure, but these lavas which Oliver places in his Grey Knotts Group vary much from flow to flow, some being porphyritic and some not. Several more outcrops of the same lava follow before we come to the next thing of note, a large erratic block built into the wall. Erratics and perched blocks are a marked feature of Langstrath all along its length, for it was heavily glaciated during the Ice Age.

We now pass through a gate, which we must not forget to close, where there are many holly trees and further exposures of lava. Notice field gates in the wall on the left, opposite the second of which a volcanic tuff is exposed. Next we come to a reminder of what our streams can do in times of flood, for debris with boulders is spread out in a huge fan. More recently a walled course has been made through the rubble and about sixty paces further on a large isolated block of flow brecciated lava stands by the wall. Through another gate a marked river terrace can be seen on our left, the top of which has been planted with trees. Our way now winds round two bends to where, by the right of the path, stands a hawthorn tree. Near this we have a porphyritic andesite, the crystals in which are of augite, a mineral with which you ought by now to be quite familiar. Look out for a big birch tree on the right, with just beyond an exposure in the floor of the track. This is a very lovely volcanic tuff and perhaps far more readily recognised than the last. Passing the old farm buildings you can see where the river once made a wide swing leaving a marked terrace to show its ancient path.

There is little worthy of comment now until we reach Galleny Force (272131), and here we seem to be on controversial ground for different workers have given differing accounts of the exposures. A fault runs down the beck here and shifts the beds on the right bank some 200 yards downstream. This fault runs well up Greenup Gill and may well have provided the line of weakness along which the valley has been eroded. By the fall itself exposures show a dark fine grained rock which has been variously described as a diabase and by other workers as a lithic tuff. In the field there is no way of settling such an argument and indeed the rocks of the Borrowdales exposed all the way up this long and lovely valley can only be regarded as an awkward squad! For myself, being no expert, I can only point out the things which interest me, and which I hope you may find of interest too. That you should always agree with my "field approximations", as the experts would say, I can neither hope nor expect. From this fall for a long way up the valley R. L. Oliver has mapped the rocks as "Dacites, rhyo-dacites, rhyolites and welded tuffs; undifferentiated," which is another way of saying that it is practically impossible accurately to record on a map the bewildering intermixtures of these many rock types!

However, this fine grained grey rock—call it what you will—

is exposed for some twenty yards or so along the stream, and then, near a bank of boulders, a different rock shows. Tough and flinty with an appearance of banding it looks to me rather like a rhyolite. But there is much of interest hereabouts besides the rocks beneath our feet, for nearby at the junction of Langstrath and Greenup becks there is an island, Smithymire (274130). We can find a clue of the meaning of the name by following the fence, for close to the stile that leads across the beck, quite plain to be seen, is an old charcoal hearth. At the head of Langstrath stands the Bowfell range and the gap between Hanging Knott and Esk Pike is known as Ore Gap. The gap is so called because the ground is red with the famous Cumberland haematite, and in the old days ore was brought down from here to be smelted in charcoal fired "bloomeries" in Langstrath to produce our first crude bits of that valuable metal, iron. Cross the stile and a little search in the bank will show bits of slag from these old bloomeries.

Notice that looking up Greenup on the right bank above the stream there is a long mound marked by parallel ridges, which is a lateral moraine. Such moraines are not uncommon in this valley and some of them form an interesting series further up the dale about the junction between Stake Beck and the main valley. They are worthy of note if you are ever so far up, but it is a little too far for today.

Continuing now by the side of Langstrath beck we come to where the stream runs down a bedding plane and, while these rocks are much easier of access from the other bank, exploration will show some of the beds to be tuff. Above the level of the fall there is a holly tree on the right with just beyond a sycamore standing on a glaciated exposure. Breaking off a piece you will find a grey tough rock which has a distinctly streaky appearance. Under the trees the rocks show marked bedding, and these are "welded tuffs"—of which more anon. Through a gate, some rough track leads to a bridge across the stream, with rather unusually a gate at the other end. This of course is to keep the sheep from crossing, so we must not neglect to close and bar it after us. Exposures here (272126) in either bank show a smooth grey glassy rock with some flow banding. Containing the occasional garnet, it is rather an attractive rock and I should think either a dacite or a rhyo-dacite.

Turning now upstream the first exposure, about 80 paces along the bank, shows a really fine welded tuff or "ignimbrite" with almadine garnets. Noted by many of the old workers on Lake District rocks and described simply as "streaky rocks", we had for long not even a clue as to their origin until a dim glint of light came out of dire tragedy. In 1902, on the island of Martinique in the West Indies, the long dormant volcano of

Monte Pelée erupted, and in a few minutes wiped out the town of St. Pierre and its 30,000 inhabitants. One marked feature of this awful disaster was the sweeping down the mountain side, with the speed of an express train, of clouds of white-hot dust. Known now as nueés-ardentes these sheets of highly mobile incandescent lava fragments settled down to form rocks of the type we now call ignimbrites, but which for long had merely been called "streaky rocks". Long and careful research in fields as far apart as New Zealand and the Valley of Ten Thousand Smokes in Alaska finally convinced vulcanologists that this terrifying phenomenon had been with us for a very long time, and that many great sheets of our ancient volcanic rocks, hitherto thought to be some form of lava, were in fact formed in this way.

Nor does this exhaust the interest of these rocks for close examination will show many small garnets, and these too have given rise to long controversy. Either they are original, and were blasted out of the vent along with the igneous material that formed this rock in which we now find them, or they are second-ary in origin having been formed subsequently within the rock by metamorphism. Personally I lean towards the first of these explanations for, having seen much of garnet bearing rocks both from at home and abroad, our local rocks do not seem to have suffered metamorphic change to anywhere near the required extent.

To tap and explore all the multitudinous exposures we shall pass now would take a very long time indeed, and in my exper-ience would give rise to a great deal of head scratching! Erratic blocks are becoming increasingly common, and passing one of these we come to a gate under the crags. Exposures here show banded rhyolite. Just beyond this gate there is a larch tree over the wall, and some 100 yards further along there is a big breast of rock on the left which rises to about 20 feet in height. At the beginning of this exposure a small stream runs down the rock face, but it tends to be somewhat intermittent. When the stream *is* running the face shows a fine ignimbrite with both large enclosed tephra and garnets, but when the water has dried up you will need to look with great care to see what a wet surface shows so well.

The track now follows first a wall and then a wire fence, with the next exposure of interest both beneath our feet and indeed all around as we thread our way among flow-banded rhyolites. One exposure shows much quartz and nearby fine flow-brecciated surfaces. Some of these surfaces have weathered into little knobbles where the fragments have been harder than the matrix. A sharp look out, however, will show exposures where the reverse has been the case giving a reticulated surface.

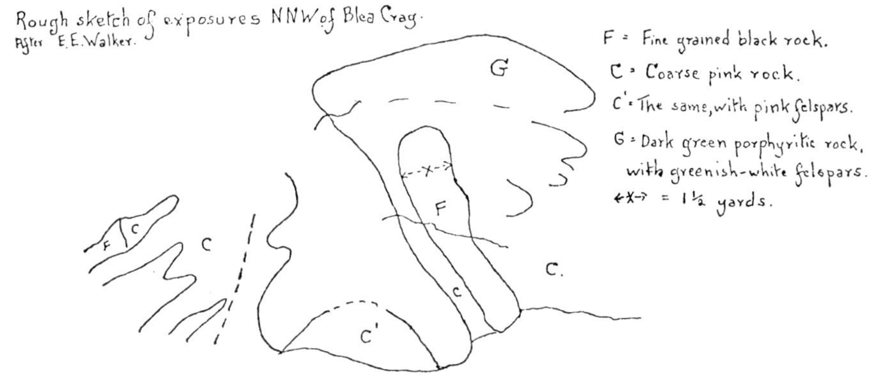

Notice now a fence which runs down at right angles from the path to where the stream leaves a really fine gorge. This is Black Moss Pot (267113), and the gorge has been cut through the hardest of our volcanic rocks by the heavily grit charged waters escaping from a post-glacial lake higher up the valley. In and around our Lakeland we have much evidence of the really tremendous erosion, wrought in a comparatively short time geologically speaking, by the melt-waters of our disappearing glaciers, charged as they were with detritus. Black Moss Pot, along with Birks Bridge in Dunnerdale, are both fine examples.

Immediately before us rises Lamper Crag and right under the big block on the end of it you can see some fine flow-banding. But now there is much evidence of the ice in moraines about the valley floor, and skirting one of these we see the object of our visit across a tangled piece of ground — Blae Crag (268106). Referred to geologically as a "complex", it consists of rocks of a widely varying texture and composition, from a dark basic variety to a light coloured granophyre, and these often grade one into the other. As we make our way towards the rock notice a rather big heather crowned erratic. It is composed of one of the light coloured porphyritic types, while just beyond the stream has cut through the drift to expose rocks which prove to be variants of the rock of the erratic block. Of the many differing types there is a dark fine-grained basic one which is well exposed on the summit of the crag. Described as a diabase one has to admit it looks not unlike an ash. There is also a dark green rock described as a quartz diabase and this displays both augites and some felspars. To complete this medley we have a coarse pink rock and a somewhat similar rock but with marked pink felspars—a complex indeed!

Blae Crag is usually noted as an example of what is known as "magmatic differentiation", a highly complex and not too well understood phenomenon. In the formation of many igneous masses there seems to be a tendency for a certain amount of sorting out of magmatic types to take place deep down in the magma chamber, but just how this is accomplished is still, and always has been, a matter of fierce argument. Nevertheless, however the actual differentiation is brought about one of the results to be seen around big igneous masses is the presence of varying rock types, and this can be seen at Blae Crag. It would seem that here, as in many other places, the earliest rocks to be emplaced tend to be of a basic character. On the Crag these must have been closely followed by felsitic types, for all around one can find exposures showing an intermingling of the two rocks, with variants of each. For such intermingling to take place very little time indeed can have elapsed between the formation and injection of the differing types. If you now care to climb the Crag you will find the diabase on the summit, while around the faces of the Crag, but perhaps more particularly on the southern face, you can find examples of both felsite veining the diabase and intermingling of basic types with the felsite.

After your explorations come back to the big heather crowned erratic for from here I can guide you to an exposure first mentioned by E. E. Walker in 1902. He says of it: "there is an exposure N.N.W. of Blae Crag ... across two streams." I brought you back to our erratic block because from here, if you look towards the north-east, you will see some 75 yards away a rocky bluff which is Mr. Walker's exposure. Although I include his original little sketch map, I have to admit that in the intervening years nature, with both grass and moss, has done a pretty effective job of camouflage. Might I suggest that if some young active party is looking for a "project" here is one ready to hand! Some stiff brushes, some buckets to carry water, and an afternoon of hard work could no doubt bring back to easy recognition what the pioneers of our Lakeland geology saw more than half a century ago!

There is little more that I can do now except, perhaps, to add that here in miniature you have what has undoubtedly taken place on a large scale in many parts of the Lake District. Around the Buttermere and Ennerdale granophyre there is a widespread and distinctly basic phase, and this seems to have preceded the granophyre itself. In fact in Wastdale there is a small outcrop, an offshoot from the granophyre, which is almost a gabbro. The same chain of events seems equally applicable to the Carrock Fell complex, as well as to the Shap Granite.

*   *   *

In conclusion may I say I have tried to give my reader a picture of Lakeland geology over the years and how various workers have tried to resolve the many problems that beset the subject. Where I have offered opinions of my own it must be remembered that they are the ideas of a completely amateur geologist whose only excuse for them is a deep and lasting interest in the subject, and our lovely Lakeland, that has never left me all the days of my life. No one in a subject like geology can claim infallibility, even more so in a complex area like ours. What I do ask is that you will try and visit each area described and see for yourself such evidence as is available, remembering too that a second visit is often more illuminating than a first!

One thing is quite certain. What you see and what you can get from what you see is dependent to a large extent on experience, and that comes and can only come from following the dictum of the great French geologist Nicolas Desmarest—"Go and see!" If my humble efforts at guidance can bring pleasure and enjoyment as well as a little understanding of a great field of human endeavour on a none too simple subject I shall be well content.

Index